MOLECULAR CLONING
OF
RECOMBINANT DNA

MIAMI WINTER SYMPOSIA

1. *W. J. Whelan and J. Schultz, editors:* HOMOLOGIES IN ENZYMES AND METABOLIC PATHWAYS *and* METABOLIC ALTERATIONS IN CANCER, 1970
2. *D. W. Ribbons, J. F. Woessner, Jr., and J. Schultz, editors:* NUCLEIC ACID-PROTEIN INTERACTIONS *and* NUCLEIC ACID SYNTHESIS IN VIRAL INFECTION, 1971
3. *J. F. Woessner, Jr. and F. Huijing, editors:* THE MOLECULAR BASIS OF BIOLOGICAL TRANSPORT, 1972
4. *J. Schultz and B. F. Cameron, editors:* THE MOLECULAR BASIS OF ELECTRON TRANSPORT, 1972
5. *F. Huijing and E. Y. C. Lee, editors:* PROTEIN PHOSPHORYLATION IN CONTROL MECHANISMS, 1973
6. *J. Schultz and H. G. Gratzner, editors:* THE ROLE OF CYCLIC NUCLEOTIDES IN CARCINOGENESIS, 1973
7. *E. Y. C. Lee and E. E. Smith, editors:* BIOLOGY AND CHEMISTRY OF EUCARYOTIC CELL SURFACES, 1974
8. *J. Schultz and R. Block, editors:* MEMBRANE TRANSFORMATION IN NEOPLASIA, 1974
9. *E. E. Smith and D. W. Ribbons, editors:* MOLECULAR APPROACHES TO IMMUNOLOGY, 1975
10. *J. Schultz and R. C. Leif, editors:* CRITICAL FACTORS IN CANCER IMMUNOLOGY, 1975
11. *D. W. Ribbons and K. Brew, editors:* PROTEOLYSIS AND PHYSIOLOGICAL REGULATION, 1976
12. *J. Schultz and F. Ahmad, editors:* CANCER ENZYMOLOGY, 1976
13. *W. A. Scott and R. Werner, editors:* MOLECULAR CLONING OF RECOMBINANT DNA, 1977
14. *J. Schultz and Z. Brada, editors:* GENETIC MANIPULATION AS IT AFFECTS THE CANCER PROBLEM, 1977

MIAMI WINTER SYMPOSIA — VOLUME 13

MOLECULAR CLONING OF RECOMBINANT DNA

edited by

W. A. Scott

R. Werner

Department of Biochemistry
University of Miami School of Medicine
Miami, Florida

Proceedings of the Miami Winter Symposia, January 1977
Sponsored by The Department of Biochemistry, University of Miami,
School of Medicine, Miami, Florida

Academic Press, Inc. New York San Francisco London 1977
A Subsidiary of Harcourt Brace Jovanovich, Publishers

Academic Press Rapid Manuscript Reproduction

ACADEMIC PRESS, INC.
111 Fifth Avenue, New York, New York 10003

United Kingdom Edition published by
ACADEMIC PRESS, INC. (LONDON) LTD.
24/28 Oval Road, London NW1

Library of Congress Cataloging in Publication Data

Molecular cloning of recombinant DNA.

(Miami winter symposia; v. 13)
Includes bibliographical references.
1. Molecular cloning–Congresses. 2. Genetic recombination–Congresses. I. Scott, William Addison, Date II. Werner, Rudolf, Date III. Miami, University of, Coral Gables, Fla. Dept. of Biochemistry. IV. Series.
QH442.M64 574.8′732 77-24886
ISBN 0-12-634250-4

PRINTED IN THE UNITED STATES OF AMERICA

CONTENTS

Poster Session: Communications

SPEAKERS, CHAIRMEN, AND DISCUSSANTS

F. Ausubel, Biological Laboratories, Harvard University, Cambridge, Massachusetts.

P. Berg, Department of Biochemistry, Stanford University Medical Center, Stanford, California.

F. Blasi, Centro di Endocrinologia ed Oncologia, Naples, Italy.

F. Blattner, Department of Genetics, University of Wisconsin Medical School, Madison, Wisconsin.

A. P. Bollon, Health Science Center at Dallas, University of Texas, Dallas, Texas.

T. W. Borum, The Wistar Institute, Philadelphia, Pennsylvania.

H. W. Boyer, Department of Biochemistry and Biophysics, University of California, San Francisco, California.

J. Carbon, Department of Biological Sciences, University of California, Santa Barbara, California.

S. N. Cohen, Department of Medicine, Stanford University Medical Center, Stanford, California.

R. Curtiss III, Department of Microbiology, University of Alabama in Birmingham, Birmingham, Alabama.

R. W. Davis, Department of Biochemistry, Stanford University Medical Center, Stanford, California.

P. Duesberg, Department of Molecular Biology, Wendell M. Stanley Hall, University of California, Berkeley, California.

A. Eisenstark, College of Arts and Sciences, University of Missouri, Columbia, Missouri.

R. Faust, USDA ARS Insect Pathology Laboratory, Beltsville, Maryland.

L. P. Gage, Department of Cellular Biology, Roche Institute of Molecular Biology, Nutley, New Jersey.

S. B. Greer (Session Chairman), Department of Microbiology, University of Miami School of Medicine, Miami, Florida.

B. Gronenborn, Institute of Genetics, University of Cologne, Cologne, Federal Republic of Germany.

B. S. Hartley (Session Chairman), Biochemistry Department, Imperial College, London, England.

L. Koditscheck, Montclair State College, Upper Montclair, New Jersey.

L. Krueger, National Institutes of Health, Bethesda, Maryland.

R. H. Lawrence, Corporate Research Laboratory, Union Carbide Corporation, Tarrytown, New York.

P. Leder, Laboratory of Molecular Genetics, National Institute of Health, Bethesda, Maryland.

J. T. Lis, Department of Biochemistry, Stanford University Medical Center, Stanford, California.

U. Z. Littauer (Session Chairman), The Children's Hospital of Philadelphia, Philadelphia, Pennsylvania.

B. Mach, Department of Pathology, University of Geneva, Geneva, Switzerland.

T. Maniatis, Cold Spring Harbor Laboratory, Cold Spring Harbor, New York.

L. Maturin, Department of Microbiology, University of Albama in Birmingham, Birmingham, Alabama.

G. M. McCorkle, Department of Biological Sciences, Purdue University, West Lafayette, Indiana.

R. B. Meagher, Department of Microbiology and Biochemistry, University of California, San Francisco, California.

J. Messing, Munich, Germany.

J. F. Morrow, Department of Biological Chemistry, Harvard Medical School, Boston, Massachusetts.

K. Murray, Department of Molecular Biology, University of Edinburgh, Edinburgh, Scotland.

O. Polianovski, Academy of Sciences, Moscow, USSR.

A. Prozorov, The Rockefeller University, New York, New York.

P. H. Roy, Biomedical Sciences, Brown University, Providence, Rhode Island.

K. Sakaguchi, Mitsubishi-Kasei Institute for Life Sciences, Tokyo, Japan.

W. Salser, Biology Department, University of California, Los Angeles, California.

P. Sarin, National Institutes of Health, Bethesda, Maryland.

W. A. Scott (Session Chairman), Department of Biochemistry, University of Miami School of Medicine, Miami, Florida.

H. K. Stanford (Session Chairman), President of the University of Miami School of Medicine, Coral Gables, Florida.

W. Szybalski, McArdel Laboratory, University of Wisconsin, Madison, Wisconsin.

J. Tooze (Session Chairman), E.M.B.O., Federal Republic of Germany.

B. Weisblum, Department of Pharmacology, University of Wisconsin, Madison, Wisconsin.

C. Weissmann, Universitat Zurich, Zurich, Switzerland.

R. Werner (Session Chairman), Department of Biochemistry, University of Miami School of Medicine, Miami, Florida.

S. Woo, Department of Cell Biology, Baylor College of Medicine, Houston, Texas.

R. Wu, Cornell University, Ithaca, New York.

PREFACE

This volume is the thirteenth of a continuing series published under the title "Miami Winter Symposia." In January 1969, the Department of Biochemistry of the University of Miami and the University-affiliated Papanicolaou Cancer Research Institute joined in sponsoring and presenting two symposia on biochemical topics as an annual event, now in its ninth year. The proceedings of the two symposia have been published as separate volumes.

As topics for the Miami Winter Symposia we select areas of biochemistry in which recent progress offers new insights into the molecular basis of biological phenomena. The first symposium sponsored by the Department of Biochemistry, emphasizes basic research in this area while the second symposium, sponsored by the Papanicolaou Cancer Research Institute, deals with the application of this research to the cancer problem. This volume contains the proceedings of the Biochemistry Department's Symposium on "Molecular Cloning of Recombinant DNA" and will be published simultaneously with the proceedings of the Papanicolaou Cancer Research Institute's Symposium on "Genetic Manipulation as it Affects the Cancer Problem" (Volume 14). Together these symposia describe some of the remarkable progress that has taken place, during the last few years, in nucleic acid technology and in its application toward analysis of the genetic organization of eukaryotic chromosomes.

Associated with the symposia is the Feodor Lynen Lecture, named in honor of the Department of Biochemistry's distinguished Visiting Professor. Past speakers have been George Wald, Arthur Kornberg, Harland G. Wood, Earl W. Sutherland, Jr., Luis F. Leloir, Gerald M. Edelman, and A. H. T. Theorell. This year the Lynen Lecture was delivered by Paul Berg. These lectures have provided insights of the history of discovery, and have included personal and scientific philosophies of our distinguished speakers. The Lynen Lecturer for 1978 will be James D. Watson and the symposia will focus on developmental biology.

To bring forward as much of the recent work as possible, short communications are presented in a joint poster session for the two symposia. Some abstracts of these short communications appear in this volume and the remainder are published in Volume 14 of the series.

Our arrangement with the publishers is to achieve rapid publication of these symposia and we thank the speakers for their prompt submission of manuscripts and Sandra Black for her efforts which enabled us to bring this about. Our thanks also go to the participants whose interest and discussions provided the interactions that bring a symposium to life and to the many local helpers, faculty and administrative staff who have contributed to the success of the present symposium. Special gratitude should be accorded to the organizers and coordinators of the program: W. J. Whelan, Sandra Black, and Olga F. Lopez.

The financial assistance of several departments in the University of Miami School of Medicine, namely, Anesthesiology, Dermatology, Pathology, Radiology, as well as the Howard Hughes Medical Institute, Abbott Laboratories, Boehringer Mannheim Corporation, Eli Lilly and Company, Hoffmann-La Roche, SmithKline Corporation, and Upjohn Company, is gratefully acknowledged.

W. A. Scott
R. Werner

The Eighth Feodor Lynen Lecture:

Biochemical Pastimes ... and Future Times

Paul Berg
Department of Biochemistry
Stanford University

The invitation to be the eighth Feodor Lynen lecturer is, for me, a great honor; furthermore, it is a very real privilege to help honor Fitzie Lynen's many outstanding scientific achievements. I am also unashamedly flattered at having my name added to the list of distinguished lecturers who preceded me, particularly as many of them were the heroes who made growing up in science such a great adventure. Two earlier Lynen lecturers warrant special mention: Harland Wood and Arthur Kornberg taught me more than the facts, concepts and skills of biochemistry; their deep commitment to learning and discovery were most influential in shaping my own goals and values in science and life. I am most fortunate now to count them as dear and cherished friends.

The Lynen lectures are traditionaly autobiographical and anecdotal. Some of the lecturers embraced this format enthusiastically (e.g. a series of "My Life and..." essays), others approached the assignment more gingerly and with trepidation (e.g. "I Hate to Bore People With My Recollections"), and one rejected it in favor of an erudite analysis of the mysterious process of creation and discovery in science. Whatever the format or style, the results have been a voyeur's delight; each offering has provided a glimpse of a remarkable man and illuminated the personal and scientific philosophies that shaped a golden era of biochemical sciences.

Partly by choice, but also because opportunities have been lacking, I have rarely (actually just once (1)) engaged in autobiographical musings. I suspect I'm not alone in this regard. Carl Cori, in a recent essay honoring Severo Ochoa (2), lamented that "the incredibly fast advance in many areas of biochemistry and molecular biology makes for equally rapid obsolescence of previous findings. Even the basic observation on which a new advance is based is rapidly forgotten because it has become common knowledge." Cori noted further that "what may be irretrievably lost in the natural course of events is... the passion, the art, the very flavor which characterizes a particular scientific period;... that, quickly sinks into oblivion together with the men and women who were the participants."

To forestall the inevitability of sinking into oblivion, I have selected several early scientific adventures to share with you - those that are especially significant to me. But the timeliness of this week's Symposium's themes and the opportunity given my colleagues to speak on their current researches, has caused me to adopt a modified oblivion protection plan; one which ensures that my present activities are also recorded in this volume. Moreover, it seems fitting at a meeting on recombinant DNA molecules for me to present a hybrid Lynen lecture.

Becoming a Biochemist: The Beginning

After a short-lived venture into chemical engineering at City and Brooklyn Colleges (now CUNY), I discovered biochemistry; the word itself (I knew little of its subject matter) appeared to offer a route to bridging my curiosity of inanimate and animate chemistry. But biochemistry at Penn State during the 1940's was concerned with agricultural, food and medicinal chemistry and my eagerness to learn about the chemical intricacies of metabolic processes went unsatisfied. To meet a requirement for a seminar course during my senior year, I submitted a paper on the then novel uses of radioactive and stable isotopes to trace the reactions of metabolic pathways. Two articles in the same issue of the 1946 Physiological Reviews, one by Buchanan and Hastings (3) and the other by Harland Wood (4), provided a wealth of examples on the use of ^{13}C- and ^{14}C- labeled compounds in the analysis of carbohydrate synthesis and breakdown. Wood's work, particularly, made a lasting impression on me. But where was Western Reserve University, the institution to which he had recently moved? (I recall thinking that Western Reserve must be an Indian reservation). R.A. Dutcher and H.O. Triebold, two professors who had read my paper and knew about my excitement of Wood's work, called to my attention an advertisement in the Chemical and Engineering News for a research assistantship in Clinical Biochemistry at Western Reserve University. But I was turned down by WRU and so accepted a fellowship at Oklahoma A and M to study chemistry. Just before leaving State College, I received a telegram offering me the assistantship in Cleveland and with some encouragement from Dutcher I changed my plans. It was a great disappointment to me to discover, after I arrived at WRU, that it was not Wood's Biochemistry department that had accepted me; instead, I was to be an assistant to Victor Myers, a distinguished clinical biochemist during the 1920's and 30's. Professor Myers asked me to analyze the cholesterol content of 75-100 post-mortem hearts and to determine if this parameter could be correlated with the cause of death or other underlying pathology.

Within a month or two of my arrival Professor Myers died and I took up with two of his former students, Jack Leonards and Leonard Skeggs. They had invited me to help them develop and exploit an "artificial kidney" machine, that they had invented. But doing nearly two years of surgery, blood and urine chemistry and experimental pathology did not distract me from the experiments in enzymology and intermediary metabolism going on in Biochemistry on the floor above. Sensing that I was at a crossroad, and with some encouragment from Warwick Sakami, I approached Wood about continuing graduate work in his department. A few members of the Biochemistry Department seemed concerned about admitting me as a graduate student; there was also some reticence by several others on the Faculty about another applicant, Jerard Hurwitz, who was then a technician in the Biochemistry Department. The controversy was resolved by a compromise; both "blackballs" were lifted and the two of us were taken in as Ph.D. students.

Although I only moved from one floor to the next, the change altered my life: I was brought into contact with people who loved and lived for biochemistry and thereby created an environment where that spark could be nurtured in others. Enzymology, intermediary metabolism and the trials of learning how to do a meaningful experiment occupied my waking hours. I was fortunate to be able to work with Warwick Sakami, Arnold Welch and Bob Greenberg on the role of folic acid and vitamin B_{12} in the biosynthesis of "labile" methyl groups and the metabolic interconversions of C1 units; these were areas of biochemistry that had only recently come to light and were gaining in interest and importance. And also, because we shared the same laboratory, late hours, an enthusiasm for research and football, as well as adjoining apartments, my wife, Millie, and I became lifelong friends with Jerry and Muriel Hurwitz.

During my stay in Cleveland there were innumerable seminars by some of the most prominent biochemists of that time. Many of them are indelibly fixed in my memory but two were especially influential: Herman Kalckar spoke (in his own inimicable way) about his use of differential spectrophotometry and enzymes to study nucleotide interconversions; Arthur Kornberg lectured on coenzyme structure and biosynthesis, emphasizing mechanisms and the use of enzymes as analytic reagents. I made up my mind then to go to each of their laboratories and learn more about enzymes and nucleotide metabolism. The American Cancer Society was very understanding and provided me with a fellowship to spend one year with Kalckar at the Institute of Cytophysiology in Copenhagen and a second year with Kornberg at the NIH.

The "Old World" Experience

The year (1952-53) in Copenhagen was idyllic. Now is not the time to reminisce about the innumerable people we met and knew, or of the social and cultural life in that beautiful city. That is for a future occasion. But I would like to recount an interesting bit of history that has never found its way into the scientific literature.

When I arrived in Denmark, during October 1952, Herman Kalckar's laboratory was deeply involved with the "galactowaldenase" system that Leloir had described a year or two earlier. While the problems that were being investigated interested me, a curious hypothesis by Thomas Rosenberg, a Danish physical chemist friend of Herman's, intrigued me more. Rosenberg suggested that the hexokinase-mediated transphosphorylation from ATP to glucose formed glucose 6-metaphosphate, which subsequently underwent a spontaneous hydration to form the well known product glucose-6 phosphate (Fig. 1A). Rosenberg and Kalckar speculated that the spontaneous (uncatalyzed) hydration of the putative glucose 6-metaphosphate might be rate-limiting; conceivably, they argued, as is the case in the hydration of CO_2, the conversion of the metaphosphate intermediate to the glucose phosphomonoester might also be catalyzed and possibly subject to regulation. There was a report that adrenal and pituitary factors as well as insulin affected hexokinase (5). Rosenberg and Kalckar wondered whether these hormones could be acting at the hypothetical glucose-6-metaphosphate hydrase e.g., whether insulin stimulated this reaction. Since it seemed reasonable that the putative metaphosphate intermediate was "high-energy" and, therefore, able to react with ADP to reform ATP, I suggested that hexokinase might catalyze a glucose dependent transfer of the terminal phosphate of ATP to IDP (Fig. 1B).

While I was preparing the labeled reagents and chromatographic methods to test our proposition, Bill Joklik came to the laboratory after completing his Ph.D. at Oxford, and decided to join me in testing that outlandish notion. Imagine

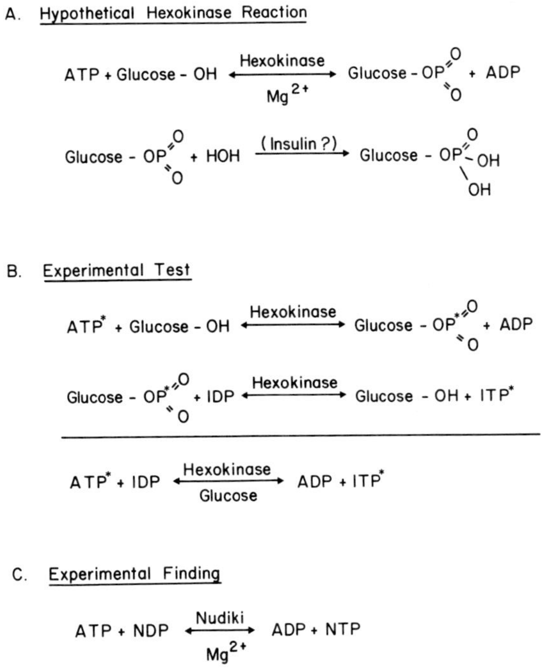

Figure 1

our surprise when the first experiment revealed a very rapid transfer of ^{32}P from ATP to ITP and from ITP to ATP; also our dismay when the reaction proceeded equally well without glucose (6)! It did not take long to prove that the hexokinase preparations, as well as most other enzymes we had in the laboratory at that time, were contaminated with a hitherto unknown enzyme that could transphorylate the γ - phosphoryl residue of nucleoside triphosphates (ribo and deoxy) to nucleoside diphosphates (Fig. 1C). Unbeknownst to us, Krebs and Hems (7), simultaneously discovered this enzyme in a more conventional way. This enzyme, which we named nucleoside diphosphokinase, but which Kalckar dubbed Nudiki, plays a key role in the synthesis of ribo- and deoxyribonucleoside triphosphates from their nucleoside diphosphate precursors. Steve Kuby's laboratory has since done an extensive characterization of the structure and properties of the enzyme (8). Our preoccupation with Nudiki and the concommitant success in the lab with the uridyl-diphospho-galactosyl transferase activity put off further attempts to explore Rosenberg's model.

I learned two important lessons from that experience: To be concerned with the purity of enzymes when examining their reaction mechanisms; and, that irrespective of whether the hypothesis or model being tested is correct, there is a strong likelihood of making an original finding if the experiment being done has never been done before.

A Return to Reality

Although our departure from Copenhagen was a sad one we were consoled by the fact that it had been a happy and successful year; besides, there were new things to learn and new worlds to conquer. During my year abroad Kornberg had moved from the NIH to establish a new Department of Microbiology at Washington University Medical School. When I arrived during November 1953, he eagerly showed me where I would work - one of the few refurbished laboratories in the Department - and even before unpacking my bags, we were discussing what problem I would tackle.

He suggested that I could pursue his earlier observation of the conversion of phosphoryl choline to phospholipid (9) or take on some aspect of the work he and Irving Lieberman were doing on pyrimidine nucleotide formation. But I had been intrigued by a report earlier that year (10), from Lipmann's and Lynen's laboratories, that apparently solved the mechanism of acetylCoA formation by aceto-CoA kinase. Although it was known that the condensation of acetate and CoA was accompanied by the cleavage of ATP to AMP and PP_i, the mechanism of this ternary reaction was a mystery. The joint Boston-Munich "communique" proposed three partial reactions to account for the overall process (Figure 2), because their enzyme preparation alone catalyzed the exchange of PP_i with the pyrophosphoryl group of ATP, as well as the exchange of CoA-bound acetyl groups with free acetate. What attracted me about their model was the proposal of an AMP-enzyme intermediate and particularly the possibility that analogous nucleotidyl-enzyme compounds, involving other ribo- and deoxynucleotides, might be substrates in polynucleotide synthesis.

But Kornberg was dubious about this possibility; moreover, he felt that the Lipmann-Lynen results could be attributed to contaminating enzymes that catalyze a reversible pyrophosphorolysis of NAD and FAD, thereby, incorporating isotope from PP_i into ATP. Nevertheless, because it seemed important to establish or demolish this model of acetylCoA synthesis, I persuaded him to let me try to isolate the putative adenyl-enzyme compound.

Jones, Lipmann, Hilz, and Lynen Model

$$\text{Enzyme} + \text{ATP} \xleftrightarrow{Mg^{2+}} \text{Enzyme - AMP} + PP_i$$

$$\text{Enzyme - AMP} + \text{CoSH} \longleftrightarrow \text{Enzyme - SCoA} + \text{AMP}$$

$$\text{Enzyme - SCoA} + \text{acetate} \longleftrightarrow \text{Enzyme} + \text{acetyl - SCoA}$$

$$\text{ATP} + \text{CoASH} + \text{acetate} \xleftrightarrow{Mg^{2+}} PP_i + \text{AMP} + \text{acetyl - SCoA}$$

Berg Model

$$\text{Enzyme} + \text{ATP} + \text{acetate} \xleftrightarrow{Mg^{2+}} \text{Enzyme - (acetyl - AMP)} + PP_i$$

$$\text{Enzyme - (acetyl - AMP)} + \text{CoASH} \longleftrightarrow \text{acetyl-SCoA} + \text{AMP} + \text{Enzyme}$$

$$\text{ATP} + \text{CoASH} + \text{acetate} \xleftrightarrow{Mg^{2+}} \text{acetyl-SCoA} + \text{AMP} + PP_i$$

Figure 2

To shorten the tale, a few fractionation steps enriched the aceto-CoA kinase activity nearly a hundred-fold and virtually eliminated the ATP-PP_i exchange. Kornberg's early skepticism had, in the end, been warranted, but was his explanation for the origin of the ATP-PP_i exchange correct? I knew, as the overall reaction equation predicted, that there was an exchange of PP_i and ATP when both of the other substrates, acetate and CoA were present. To determine if both acetate and CoA were essential for the ATP-PP_i exchange each one was omitted from the reaction. I was flabbergasted to find that omission of CoA increased the rate of exchange and that leaving out acetate alone eliminated the ATP-PP_i exchange completely. This observation suggested immediately that the aceto-CoA kinase catalyzed a reversible reaction between ATP and acetate resulting in the release of PP_i; the most reasonable intermediate seemed to be a hitherto unknown mixed anhydride, acetyl adenylate, strongly bound to the enzyme (11) (Fig. 2). Synthetic acetyl adenylate proved to be considerably more unstable than acetyl phosphate but it was quite straight forward to show that it yielded ATP in the presence of PP_i and acetyl CoA with CoA (11). Additional experiments proved that the Lipmann-Lynen model was incorrect and supported my formulation of the reaction (Fig. 2).

Once again experiments with an impure enzyme had given misleading information about its mechanism; in this case the most likely impurities in the enzyme preparation were free acetate and AMP. When I first reported these findings in 1955 at the ASBC meetings in San Francisco I was quite apprehensive about how they would be received by Lipmann and Lynen. But my apprehension was unfounded because Fritz Lipmann was most generous in his praise of my work and genuinely delighted at the novelty of the mechanism of the reaction. Years later, when I first met Fitzie Lynen, he was equally complementary. It was, indeed, one of the exciting chapters of my scientific career. I shall always be grateful to Kornberg for his support and encouragement during that work and for insisting that I publish the work independently.

Becoming a Molecular Biologist

During the search for the hypothetical adenyl-enzyme compound in other cell extracts, I discovered other ATP-PP_i exchange activities. Of particular interest was the finding that amino acids promoted an ATP-PP_i exchange. It seemed reasonable, by analogy, with the aceto-CoA kinase reaction, that this occurred via the formation of aminoacyl adenylates (the first reaction in Fig. 3), an assumption which proved to be correct. This was established when I purified an enzyme that catalyzed the formation of L-methionyl adenylate using ATP and L-methionine as well as the formation of ATP from synthetic methionyl adenylate and PP_i (12). Jim Ofengand, my first graduate student, and I reasoned that there must be a natural aminoacyl acceptor analogous to CoA and we set out to find the hypothetical acceptor. I thought it might be the template for protein synthesis, but the acceptor turned out to be small polynucleotides subsequently named tRNAs. (13) (The second reaction in Fig. 3).

The elegant work of Mahlon Hoagland and Paul Zamecnik (14) established that tRNAs were, in fact, the adaptors Crick foresaw (15) would be needed to translate the genetic code words for protein synthesis. Our subsequent work established that separate, specific enzymes (aminoacyl tRNA synthetases) catalyzed the attachment of each amino acid to one or a few cognate tRNA chains (see reaction 3 in Fig. 3) (16).

Aminoacyl Activation for Protein Synthesis

$$\mathrm{ATP} + \mathrm{aa_x} + \mathrm{Enzyme_x} \xleftrightarrow{Mg^{2+}} \mathrm{Enzyme} - (\mathrm{aa_x} - \mathrm{AMP}) + \mathrm{PP_i}$$

$$\mathrm{Enzyme} - (\mathrm{aa_x} - \mathrm{AMP}) + \mathrm{tRNA^x} \longleftrightarrow \mathrm{aa_x} - \mathrm{tRNA^x} + \mathrm{AMP} + \mathrm{Enzyme}$$

$$\mathrm{ATP} + \mathrm{aa_x} + \mathrm{tRNA^x} \xleftrightarrow{Mg^{2+}} \mathrm{aa_x} - \mathrm{tRNA^x} + \mathrm{AMP} + \mathrm{PP_i}$$

Figure 3

In the midst of these experiments (1959), I, along with other members of the Washington University Microbiology Department, moved to Stanford University where we established a new Department of Biochemistry. There followed a period of nearly ten years occupied with purifying and characterizing individual aminoacyl tRNA synthetases, tRNA acceptors and studying the mechanism and specificity of aminoacyl tRNA synthesis. Though this period is filled with colorful memories of people and events, I shall forego the temptation to discuss them now.

Concurrently, another enzyme and problem occupied our interest. Mike Chamberlin, for his Ph.D. thesis research, detected and purified an enzyme which synthesized RNA using DNA as a template (17) and, thereupon, followed a succession of studies concerning the mechanism of RNA synthesis. I recall vividly our excitement when Bill Wood, who was also a graduate student at the same time, found that RNA made with Chamberlin's purified RNA polymerase and T4 DNA was translated into protein with a fractionated protein synthesizing system (18). But the euphoria was shortlived when we heard of Marshall Nirenberg and Heinrich Matthaei's startling finding that poly U could direct the assembly of polyphenylalanine, in vitro (19). Only more recently has the coupling of DNA transcription to protein synthesis in vitro come into its own in studies of the mechanism of regulation of gene expression.

Dissecting and Reconstructing the SV40 Chromosome

So much for my recollections of the past. What of our present efforts at analyzing the mechanism of eukaryote gene expression? Sometime during 1965-66 I became acquainted with Renato Dulbecco's work on polyoma virus. The growing sophistication of animal cell culture methods, the ability to follow and quantify viral multiplication in cultured animal cells and the discovery of polyoma-mediated oncogenesis in vitro captured my attention. Expecially intriguing was the apparent similarity polyoma-mediated oncogenesis had with bacteriophage - induced lysogeny. I felt that this system could provide an experimental handle for studying the mechanism of eukaryote gene expression and its regulation. It seemed worth a sabbatical year to find out.

During 1967-68 I joined Dulbecco's laboratory at the newly opened Salk Institute in La Jolla, California. Under Marguerite Vogt's and Walter Eckhart's day to day expert tutelage, I, my longtime assistant Marianne Dieckmann and François Cuzin, a post-doctoral fellow from the Pasteur Institute, were transformed from biochemical prokaryotologists to animal virologists. We even managed an interesting piece of work on the formation of oligomeric forms of polyoma DNA (20). During that year, my conviction that the tumor virus system provided an interesting model for biochemical studies of gene expression in animal cells become firmer and I made plans to begin research in this area on returning to Stanford.

By 1970-71 the work on tRNA structure-function and suppression (21) was winding down and being replaced by concerns and experiments about the molecular biology of a related tumor virus, simian virus 40, or SV40. We've been almost totally committed to that pursuit ever since.

What is so interesting about SV40 and why have so many other laboratories adopted it as an experimental model?*

---

* In summarizing the salient features of the molecular biology of SV40 I have not cited individual investigators or laboratories for their contributions: but my purpose was to simplify the presentation and save time. For those who want entry to the original literature see the excellent volume by Tooze, et al., (22) and more recent reviews (25).

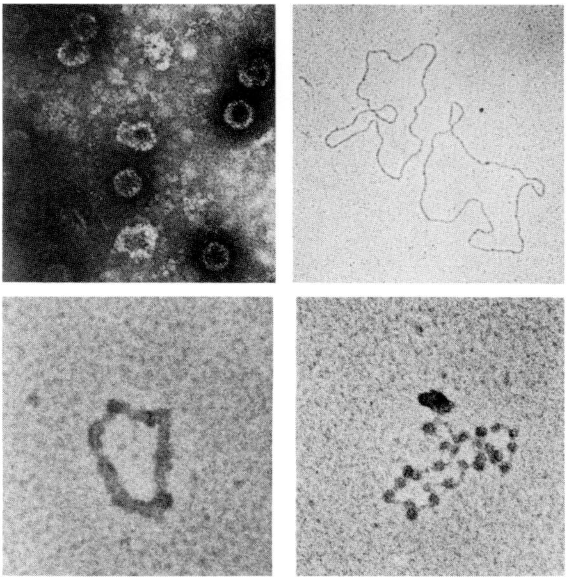

Figure 4. Electron micrographs of: u.l) SV40 virions; u.r.) SV40 DNA; l.l.) "condensed" SV40 minichromosome; l.r.) "relaxed-beaded" SV40 minichromosome. Photo by J. Griffith.

Structurally, SV40 is relatively simple: the nearly spherical particles (Fig. 4, upper left) consist of a protein "shell" containing two or three viral-coded polypeptides; enclosed within the capsid is a double-stranded circular DNA molecule of 5300 base pairs (5.3 kb) (Fig. 4, upper right), complexed with four histones, H2a, H2b, H3 and H4 and appearing as a condensed (Fig. 4, lower left) or "beaded" (Fig. 4, lower right) chromatin-like structure. The 5.3 kb DNA molecule codes for the capsid proteins and the A protein (or T antigen), which is necessary for viral DNA replication and cellular transformation. (The A protein may be the precursor of two other viral coded products - the tumor-specific antigen (TSTA) and U antigen). Having genes, histones and a chromatin-like structure qualifies SV40 as a minichromosome.

Restriction endonucleases have played a crucial part in defining the genetic and functional organization of the SV40 genome (25). The restriction or cleavage sites serve as coordinates for a molecular map of the DNA and permit one to locate, accurately, particular physical features or genetic loci. The single EcoRI endonuclease cleavage site serves as the reference marker and is assigned map position 0/1.0; all other positions in the DNA are given map coordinates in SV40 DNA fractional lengths measured clockwise from 0/1.0 (Fig. 5).

Geography of SV40 Chromosome

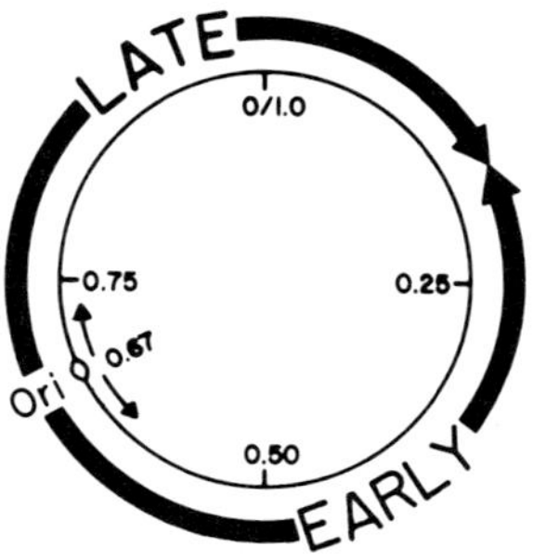

Proteins	Complementation Groups
VP1 (45K)	[B,C,BC], D,E
VP2 (35K)	
VP3 (25K)	
A (70-95K)	A
TSTA (?)	
U (?)	

Figure 5. The "early" proteins (A, TSTA and U antigen) are expressed from the "early" region, while the capsid proteins, VP1, VP2, and VP3 are coded for by the "late" region.

The SV40 chromosome, after it enters the nucleus of infected primate cells, is expressed in a regulated temporal sequence. Initially, a viral RNA transcript, complementary to one strand (the E-strand) of about one half of the DNA, appears in cytoplasmic polysomes. This transcript, synthesized in the counter clockwise direction (see Fig. 5), encodes the structural information for the A protein (and possibly TSTA and U antigens). Synthesis of A protein triggers viral DNA replication which begins at map position 0.67 and proceeds bi-directionally, terminating about 180° away at map coordinate 0.17. Concommitant with DNA replication, two, or perhaps three, new viral RNA transcripts appear in the polysomes; these transcripts, synthesized in the clockwise direction, are complementary to the other strand (the L-strand) of the other half of the DNA (see Fig. 5). These late mRNAs code for the virion proteins VP1, VP2, and VP3. Synthesis of progeny DNA molecules and the capsid proteins results in death of the cell and release of mature virions.

SV40 also has an alternative life cycle during infection of non-permissive hosts (rodent and other non-primate cells). The same early events take place (synthesis of the E-strand transcript and A protein) but DNA replication, late strand transcription and virion protein synthesis do not occur. Generally, cell DNA replication and mitosis is induced after infection but most cells revert to a normal state and show no evidence of prior infection. A small proportion of the cells (less than 10%) acquire new properties and capabilities (transformation). The transformed cells: 1) continue to divide under culture conditions that restrict the multiplication of normal cells, 2) have an altered morphology and 3) produce tumors when innoculated into appropriate animals. Moreover, the transformed cells contain all or part of the viral DNA, covalently integrated into the cell's chromosomal DNA.

When we began our work much of what I have just summarized was unknown. But Peter Tegtmeyer (26) and later, Robert Martin (27) had begun a genetic analysis of SV40. They isolated thermosensitive (ts) mutants that were defective in both DNA replication and transformation and other mutants which performed these functions normally but failed to make infectious virions. Their mutants were eventually sorted into different complementation groups (see Fig. 5), presumably corresponding to functional genes; but mapping the mutational sites and correlating the changes in the DNA sequence with alterations of viral polypeptides or control sites was not possible then. Accordingly, we set out in 1972 to isolate SV40 mutants with substantial alterations in their DNA structure (e.g. deletions, substitutions or additions, etc.) and to map these changes on the DNA molecule by physical and enzymatic means. We expected that deletions in structural genes would cause easily demonstrable changes in their polypeptide products.

Janet Mertz and I suceeded in obtaining and characterizing cloned isolates of a variety of deletion mutants that arose spontaneously (28); but most of these were only marginally useful for the purpose we had in mind because their deletions were too large and there were extensive rearrangements in the DNA. What was important about Mertz's achievement, however, was that she solved the problem of how to clone and propagate defective SV40 genomes; unlike ts mutants, deletion mutants are not conditional in their phenotype. Mertz employed the Tegtmeyer-Martin ts mutants to supply the function inactivated by the deletion. Mutants with deletions in a late gene can be cloned (as single plaques) and propagated by complementation in cells coinfected at 41°C with ts mutants defective in the early function and vice versa. (29)

For the past few years we have explored biochemical approaches to constructing deletion mutants. Lai and Nathans (30), as well as our laboratory (28), used restriction enzymes that make multiple cleavages in SV40 DNA to resect segments of the genome. But the most generally applicable and simplest method was discovered by John Carbon (31) during a recent sabbatical leave at Stanford.

Digestion of SV40 DNA by restriction or other endonucleases converts the circular DNA to full length (after 1 cleavage) or shorter linear molecules (after more than 1 cleavage). The 5'-ends of such DNA molecules can be trimmed away with phage λ exonuclease to generate molecules with 3'-ended, short, single-stranded tails (Fig. 6). Carbon discovered that cells infected with such linear DNA molecules, rejoin their ends to regenerate circular structures; but these lack between 5 to 200 nucleotides at the site of the initial cleavage. Possibly, this occurs because cells can transiently pair short homologous regions in the single-stranded tails, digest away the unpaired portions, fill in the gaps and covalently join the nicks to regenerate covalently closed, shorter, circular molcules. With a site-specific endonuclease, the deletion occurs at the cleavage site(s); with a relatively non-specific endonuclease, e.g. S1 nuclease or DNase I with Mn^{2+}, a circularly permuted set of linear DNA molecules is produced and infection with these ultimately yields mutants with deletions distributed throughout the entire genome. Viable deletion mutants can be propagated without a helper virus, whereas mutants with deletions in structural or regulatory genes generally need a complementing helper genome to grow.

For mutants with small deletions to be useful, a method is needed to map their physical location on the viral DNA. Tom Shenk, Peter Rigby and Carl Rhodes showed (32) that this

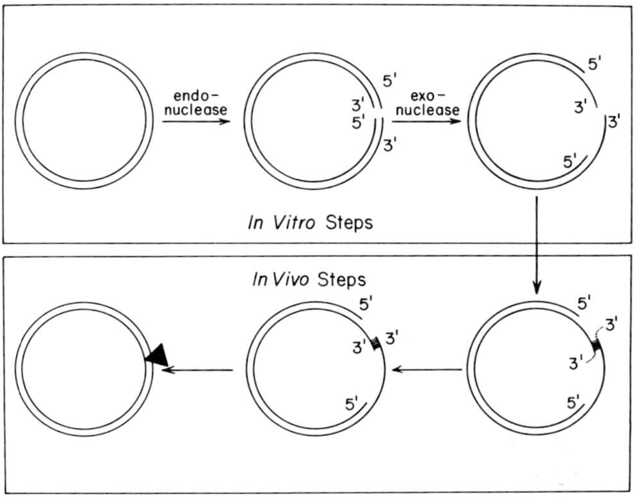

Figure 6. Biochemical and cell-mediated steps in production of small deletions in SV40 DNA. The cell-mediated pathway is hypothetical.

also could be achieved by biochemical means. When identical linear DNA molecules are denatured and renatured, the original fully duplex structure is regenerated. If two DNAs differ in a portion of their molecular sequence, heteroduplexes are produced by the denaturation and renaturation procedure. These heteroduplex molecules are double-stranded in their homologous segments and have single-stranded regions where they differ (see Fig. 7). For example, when wild type DNA and a deletion mutant are cleaved with EcoRI endonuclease to produce linear molecules, and these are denatured and renatured, heteroduplexes which contain a single-stranded loop at the site of the deletion are formed. With deletions of 200 nucleotides or more, the single-stranded loop can be visualized in the electron microscope, but small loops of 10 to 150 nucleotides are difficult or impossible to see.

The S1 nuclease from Aspergillus, which is specific for degrading single-stranded DNA, can digest away the small loop and cleave the exposed portion of the opposite DNA strand to produce two fragments whose length is determined by the position of the deletion. In this simple way, a deletion of as

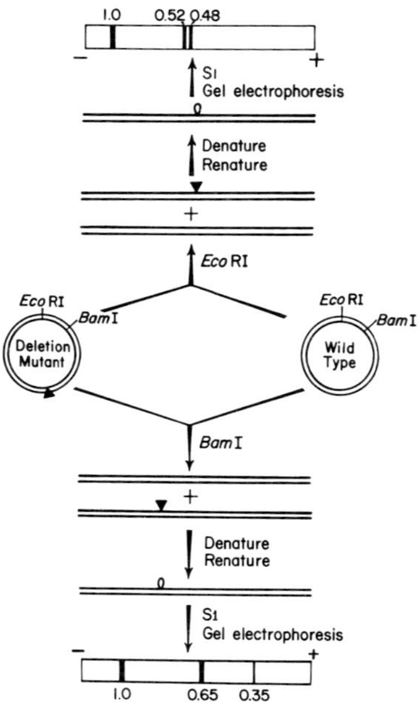

Mapping Deletion Mutations In SV40 DNA With Enzymes

Figure 7. Mapping deletion mutations in SV40 DNA with enzymes. The small triangle represents the site of the deletion in the mutant molecules, and the loop the unpaired region in the heteroduplexes. The size of the fragments produced by S1 nuclease digestion of the heteroduplexes is deduced from their electrophoretic mobility in agarose gels (diagram of gels at top and bottom).

few as 5 base pairs in SV40 DNA can be located readily with respect to the EcoRI cleavage site. If heteroduplexes are also made from linear molecules generated with another restriction enzyme, e.g. BamI, a different set of fragments is generated in the S1 nuclease digestion. Figure 7 illustrates how this procedure was used to map a deletion at map coordinate 0.48. S1 nuclease is remarkable in detecting non-homologous regions in a duplex structure; it can detect even some single base mismatches (32).

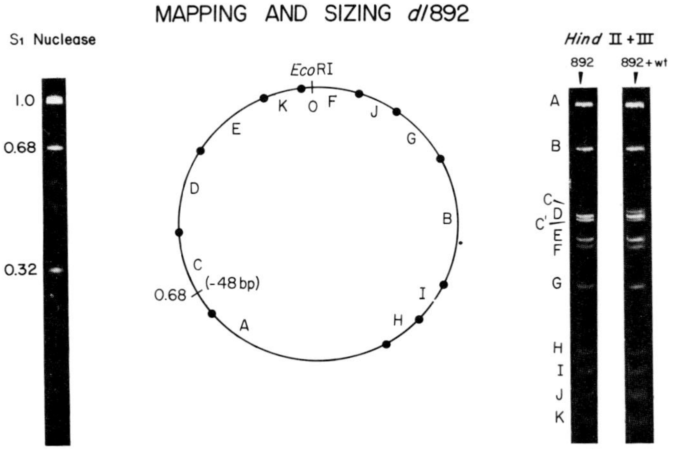

Figure 8. The center circle indicates the fragments produced by cleavage of SV40 DNA with Hind II and III endonuclease. At the left is the result of the S1 nuclease mapping analysis of a mutant with a deletion at map position 0.68. The right panels show the Hind II and III endonuclease digests of the mutant DNA and a mixture of the mutant and wild type DNA. The mutant DNA lacks the normal C-fragment and yields a faster migrating C-fragment.

Beside the map location it is necessary to determine how much of the DNA has been deleted in the mutant. This can be determined by a comparison of the electrophoretic mobility of restriction endonuclease-generated fragments from mutant and wild type DNAs. Fig. 8 shows that a mutant produced by Carbon's procedure yields an altered C-fragment after digestion with Hind II and III restriction endonucleases. Considering that the S1 nuclease mapping procedure showed that the deletion in this mutant was either at map position 0.68 or 0.32, the occurrence of an altered C-fragment (the segment bounded by coordinates 0.655 and 0.76) unequivocally fixes the deletion at 0.68. The increased mobility of the altered C-fragment is consistent with a shortening or deletion of 48 base pairs.

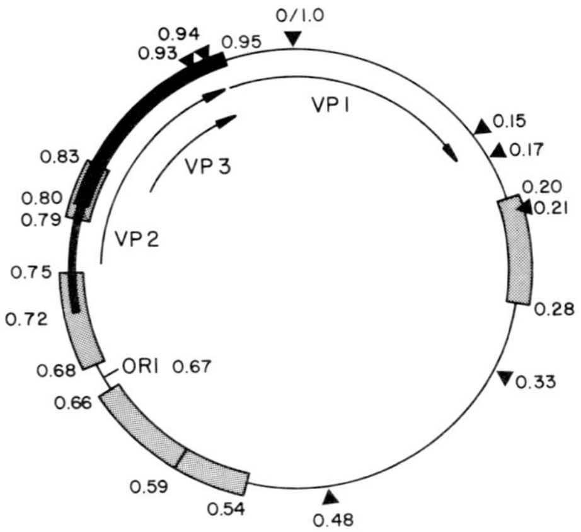

Figure 9. A summary of the deletion sites in SV40 produced at Stanford. The open and closed bars indicate extended deletions between the coordinates indicated. The solid triangles indicate the map position of small deletions (< 100 bp).

What have we learned about the genetic organization of the SV40 chromosome using these methods for constructing and mapping deletion mutants? Fig. 9 shows a map of SV40 indicating those regions within which deletions have been introduced and whose phenotype has been scored. (Cole, Landers, Goff, Manteuil-Brutlag, Dieckmann and Berg, unpublished observations.)

A somewhat unexpected discovery was that some regions of the genome are dispensible: One such region occurs between coordinates 0.17 - 0.19, another between 0.68 - 0.76 and a third, between 0.54 - 0.59. Preliminary sequence analysis (Dhar, Subramanian, and Weissman, personal communication) of a mutant in the first class suggests that nucleotides near the 3' ends of both early and late mRNAs are deleted; this apparently has no effect on the virus' ability to grow or transform non-permissive cells. Although mutants lacking sequences at map positions 0.68 - 0.75 grow more slowly than their parents, they do not have altered virion proteins and are normal in their transforming efficiency; the reasons for their slower growth is still under study, Mutants with deletions of 15 to 195 base pairs within the segment between map positions 0.54 - 0.59 also grow well. Because these deletions were thought to occur within the putative structural gene for the A protein, we and Tegtmeyer compared the electrophoretic mobility (presumably, the polypeptide molecular weight) of the A protein produced by wild type virus and by a mutant having a deletion of 184 base pairs within this region; no difference was discernible. The obvious inference that the beginning of the structural gene for the A protein is distal to map position 0.54 is, however, uncertain since the estimated size of the A protein, about 90-95 kilodaltons, is too large to be coded by the nearly 2 kb of DNA in the early region. Further work is needed to resolve this paradox.

Still incompleted experiments, carried out in collaboration with Bouck and diMayorca indicate that the mutant with the 184 base pair deletion in the 0.54 - 0.59 region has a lowered transforming activity than the wild type virus. This point is important in view of Benjamin's laboratory's finding that analogous deletion mutants of polyoma fail to transform non-permissive cells and have a reduced ability to multiply in stationary phase cells (33).

Another interesting deletion mutant lacks the segment between map coordinates 0.59 - 0.66; that mutant is A^- and fails to complement the growth of tsA or other early deletion mutants. Mutants with deletions at 0.47 - 0.48, 0.33 and 0.20 - 0.28 are also A^- (except a small deletion at 0.21); since deletions at 0.17 - 0.19 are viable and deletions that extend clockwise to 0.15 are A^+, the likely boundaries of the functional A gene has been provisionally assigned to 0.54 - 0.21. Why then is the region between 0.59 - 0.66 essential for A gene expression? Possibly, that segment contains a promoter for early gene transcription, a processing site for processing of the A gene mRNA, a ribosome binding site for translation of the A gene or some other yet unknown control region. Only time and more experiments will clarify the issue.

Deletions in the late region of the genome have also been informative (Fig. 9). Mutants having deletions at map position 0.79 - 0.83 define a new complementation group, since they help all known ts mutants to grow. Each of the mutants in this group makes normal VP1 and VP3 capsid proteins; but some fail to make VP2, others make a smaller VP2 and a few make an apparently unaltered VP2. Small deletions at 0.93 and 0.94 cause VP2 and VP3 polypeptides to be about 6 and 4 kilodaltons, respectively, smaller than the wild type proteins.

These two deletion mutants, however, do not alter the size of VP1. These findings indicate that the structural gene for VP3 lies within the map coordinates 0.83 - 0.945 and VP2 between 0.76 and 0.945.

Deletions occuring at map coordinate 0/1.0 and at 0.15 alter the size of VP1 but those at 0.17 do not. Our findings, those of Lai and Nathans (34) and particularly, those by Fiers <u>et al</u>., (35) that show the initiator codon of VP1 to be at 0.945, suggest that the limits of the VP1 structural gene are between 0.945 - 0.16.

A particularly interesting mutant is one with a deletion of the region 0.72 - 0.80. This mutant, which expresses normal A gene function, fails to complement mutants with defective VP1 and VP3 proteins, even though the genes coding for these proteins are unaltered. Evidently the region 0.76 - 0.80 contains information essential for expressing the distal genes, but the nature of that information is unknown.

Mutants with deletions close to the origin of DNA replication also shed some light on the essentiality of that structure. Fig. 10 shows the nucleotide sequence between map coordinate 0.64 - 0.17, as determined by the very elegant work of Subramanian, Dhar and others in Sherman Weissman's laboratory (36). It is a very complex region having several "true" palindromes (bp 56 to 66 and 102 to 118), perfect and nearly perfect inverted repeats (e.g. 139 to 165 and 82 to 96), a region containing exclusively A/T base pairs (bp 167 to 183), four non-tandemly repeated 8 - 9 base pair sequences and one tandem repeat of a 55 base pair segment (bp 260 to 314). Which of these sequences is required for initiation of DNA replication? Does this region contain information for initiating early (37) and/or late (38) RNA transcripts as has been suggested by several laboratories?

Proceeding from top left to lower right (map position 0.64 - 0.71), the deletion, dl 1209, which is shown as a solid line drawn between base pairs 0 - 82 (or possibly through 101), removes a palindromic sequence, several inverted repeats and a region corresponding to a 5' - terminus of late RNA. The second deletion, dl 892, removes base pairs 187 - 206 which contains all of the first and part of the second 8 - 9 base pair non-tandem repeat sequence. Tom Shenk, now at the U. of Connecticut Medical Center, has expanded this deletion in both directions so that one end probably invades the exclusively A/T segment (base pair 167 - 183) and the other extends beyond base pair 245 thereby removing the other copy of the 8 - 9 base pair non-tandem repeat. The mutant with the

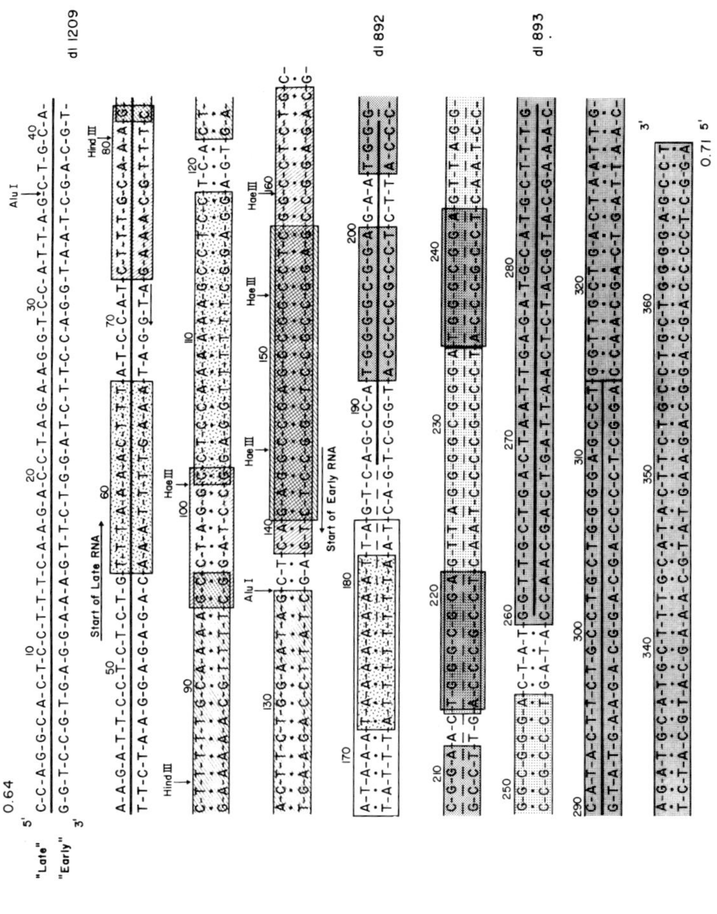

Figure 10. The nucleotide sequence of SV40 DNA from map position 0.64 to 0.71. The shaded areas indicate segments of unusual sequence arrangement.

deletion extending into the A/T segment grows poorly, but the one with the deletion extending to position 245 shows no gross abnormality in its growth (Shenk, private communication). The third deletion, dl 893, removes almost all of one of the 55 base pair tandem repeats between base pair 260 and 314 showing that this structure is not obligatory for replication. Further studies of the sequence alterations produced by deletions within this region should help clarify the structural features of a replication origin.

SV40 DNA as a Cloning Vector

Shortly after beginning work with SV40 and animal cells, I became intrigued with the notion of using SV40 as a transducing virus to introduce foreign genes into cellular chromosomes. One strategy that seemed worth exploring was to biochemically insert "foreign" DNA segments into the viral DNA and use the capacity of SV40 to integrate its DNA into the cellular genome. To pursue that approach we needed a source of genes and a means of "splicing" them into the circular SV40 DNA.

Fortunately, Douglas Berg, Dave Jackson and Janet Mertz had constructed, by conventional genetic means (39), a λdv plasmid containing the E.coli genes coding for the enzymes of galactose utilization. Dave Jackson, Bob Symons and I were then able to construct a molecular hybrid between the bacterial plasmid and SV40 DNA (40) by relatively straightforward enzymologic procedures using available purified enzymes (Fig. 11).

The circular SV40 DNA and λdvgal plasmid DNA were first cleaved with a specific endonuclease to convert them to linear molecules. Then, after a brief digestion with λ-exonuclease to remove about fifty nucleotides from the 5'-termini, short "tails" of either deoxyadenylate or deoxythymidylate residues were added to the 3'-termini with purified deoxynucleotidyl terminal transferase. The two DNAs with their complementary "tails" were joined and cyclized by simply mixing and annealing them under appropriate conditions. The gaps, occuring at regions where the two parental molecules are joined, were filled in with DNA polymerase I and the resulting molecules were covalently sealed with DNA ligase; exonuclease III repairs nicks or gaps created during the manipulations. This method, which was also developed independently by Lobban and Kaiser (41), is general and can be used to join together any two DNA molecules irrespective of their source or the structure of their termini.

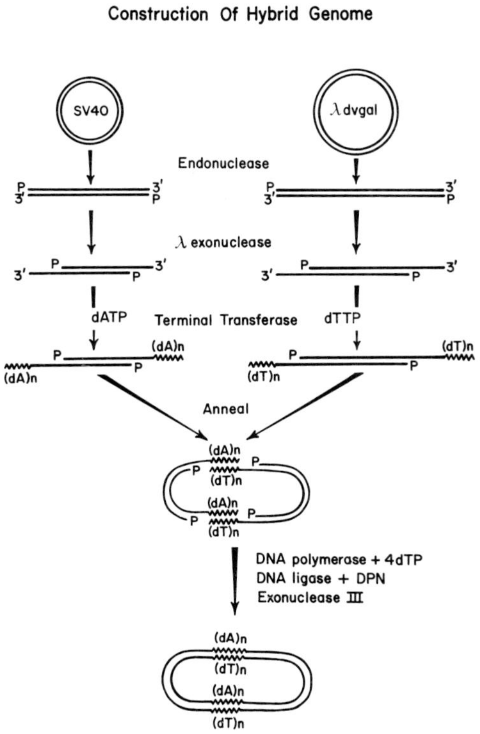

Figure 11. Construction of a hybrid genome containing genes of SV40, bacteriophage λ and E.coli.

Our goal in constructing this hybrid, or recombinant DNA as such molecules are now called, was to introduce the E.coli gal genes into the genome of animal cells to test for their expression and regulation. But we recognized that introduction of the same λdvgal-SV40 recombinant DNA molecule into E.coli also had interesting experimental ramifications. However, the latter experiment was aborted after the recombinant DNA was made, because of concern for the potential biohazard posed by accidental dissemination of such transformed E.coli.

Since that time there has been an explosive growth in the application of recombinant DNA methods to a variety of challenging problems, several of which are included in this volume. This phenomenal leap forward owes much of its impetus to the growing sophistication about the properties and use of restriction endonucleases, the development of easier ways of recombining different DNA molecules and, most importantly, the development of vector molecules, bacterial plasmids and phages, that enable recombinant DNA molecules to be established in E.coli where they can be propagated and amplified.

Cloning experiments now underway in many laboratories will soon lead to the accumulation of an extensive library of eukaryote DNA segments containing single or clusters of genes. Determining their molecular anatomy, and arrangement, will occupy many man and woman hours; undoubtedly, there will be many surprises. As our sophistication about the organization of eukaryote chromosomes increases, the theories concerning mechanisms of regulation of gene expression, development, etc., will be refined. But will we be able to deduce how these isolated genes work in the chromosomes from whence they came? Can we understand the expression of mammalian genes when they are embedded in a bacterial plasmid or phage and divorced from the intrinsic and extrinsic influences they contend with in their normal host? Even if such genes can be turned on in prokaryote hosts by ingenious genetic manipulation, will that answer our questions about how such genes are controlled in situ? Ultimately, we shall need a way to introduce the isolated eukaryote genes back into their original or related hosts to plug that gap.

Our increased understanding of the genetic organization and expression of the SV40 chromosome encouraged us to explore the use of portions of the viral DNA to clone, propagate and integrate foreign DNA segments into cultured animal cells. Accordingly, Steve Goff and I set out about 18 months ago to adapt SV40 as a cloning vector. So far, because of restrictions imposed by the NIH guidelines on recombinant DNA research (42), our efforts have been limited to recombining purified, defined segments of prokaryote DNA with the SV40 vector; moreover, the experiments have been done in the recommended fashion using approved P3 containment facilities.

Goff's first vector was designed to be propagated as a virus; consequently the total size of the recombinant molecule cannot exceed 1 SV40 DNA length. SVGT-1 which can be used to clone foreign DNA segments of 1 to 2 kb, was prepared by cleaving SV40 DNA sequentially with HpaII and BamI restriction endonucleases (Fig. 12A); the 3.2 kb segment, which contains the origin of DNA replication and the entire early region of the genome serves as the vector. The first insert was a 1.5 kb segment of λ phage DNA containing at least three transcriptional promoters (Prm, Pr, and Pre) and two structural genes (cro and cII). This segment was excised from λ DNA by

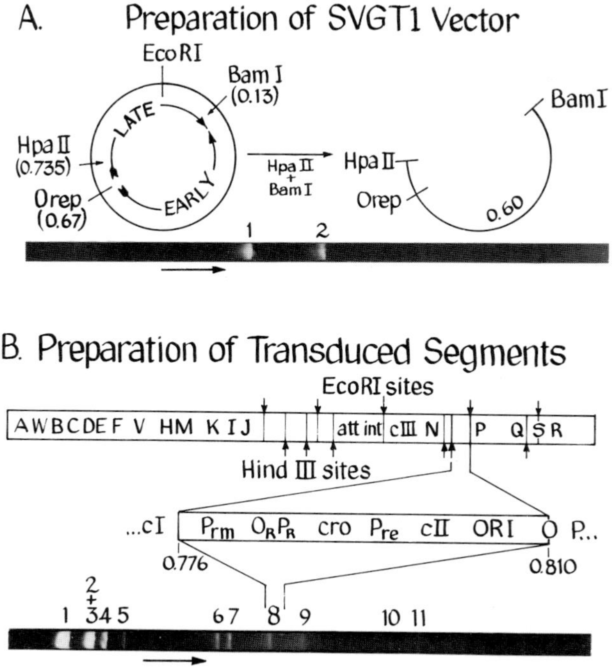

Figure 12. Protocol for Constructing λ-SVGT-1 Recombinant. A) The vector, SVGT-1, was prepared by cleaving SV40 DNA with the indicated restriction endonucleases and isolating the 3 kb segment containing the origin of DNA replication and the entire early region. Oligo dA "tails" were added to its 3' ends. B) The transduced segment, from phage λ DNA, was obtained by sequential cleavage with Hind III and EcoRI endonucleases. Oligo dT "tails" were added to its 3' ends.

successive cleavages with Hind III and EcoRI endonucleases and purified by electrophoresis (Fig. 12B). After adding poly dA termini to SVGT-1 and poly dT termini to the λ DNA segment, the samples were mixed, annealed and used to infect monkey cell cultures that were coinfected with a helper virus DNA to provide the missing late functions. About a third of the plaques had virus particles containing λ-SVGT-1 recombinant genomes, as judged by reassociation kinetic measurements, by restriction enzyme analysis of the recombinant DNA, and by electron microscopy of heteroduplexes between the hybrid and SV40 DNAs. These tests confirmed that the hybrid molecules had the predicted structure: the vector SVGT-1 DNA segment joined by short lengths of dA:dT (about 100 bp) to the 1.5 kb segment of λ DNA. As expected, from the nature of the joining process, there were nearly equal numbers of recombinants with the λ DNA segment inserted in one or the other orientation relative to the vector.

Here I want to digress from the analysis of these recombinant genomes to speak of a method Luis Villarreal and I developed (43), which greatly simplifies the search for recombinant viral genomes and simultaneously substantially reduces the risk of disseminating them. A monolayer of cells and the associated plaques, such as is produced in the standard plaque assay for titering infectious virus, can be transferred quantitatively to a nitrocellulose disc merely by laying the disc onto the cell layer (Fig. 13A). Treatment

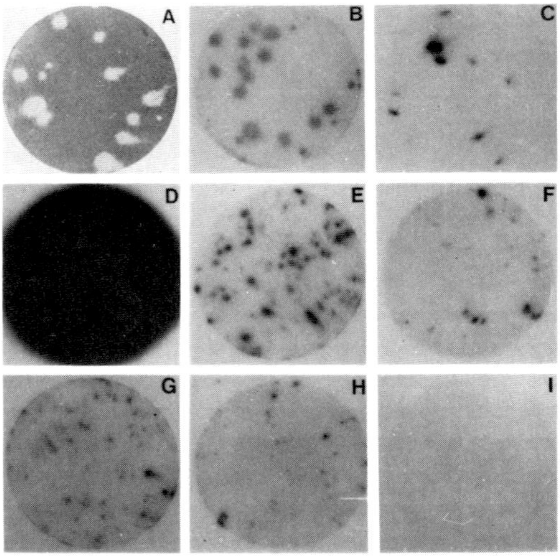

Figure 13. Images of virus plaques transferred from infected cell monolayers to nitrocellulose discs and hybridized to radioactive probes homologous to the viral DNA sequences. (See text for explanation of results.)

of this imprint of the cells and plaques with alkali, dissolves the cellular material and causes the denatured DNA to bind to the filter. Hybridization of the DNA fixed to the disc with a radioactive probe of RNA or DNA complementary to the sequence being sought, and radioautography, reveals which plaques or other locations in the monolayer contain the desired sequence. With a labeled SV40 DNA probe, plaques containing SV40,appear black on the radioautogram (Fig. 13B); with a labeled λ DNA probe, the radioautogram is blackened over plaques that contain the λ-SVGT-1 recombinant (Fig. 13C). The viral contents of a plaque can be readily recovered from the corresponding regions of the agar overlay that is removed prior to making the transfer to the nitrocellulose. Since imprints of monolayers infected with SV40 alone, do not yield positive areas with a labeled λ DNA probe (Fig. 13I), as few as 1 to 50 λ-SVGT-1 recombinants can be detected in a monolayer infected with 10^4 (Fig. 13D-F) or even 10^5 pfu (Fig. 13 G, H) of wild type virus.

The ability to detect recombinant genomes does more than simplify the task of isolation. The procedure reduces the probability of dispersing the viral recombinants, because it eliminates the preparative steps used to isolate the virus and DNA from innumerable plaques and restricts the screening operations to the more secure confines of a petri plate and nitrocellulose discs.

An interesting and important question is whether the λ DNA sequence in the λ-SVGT-1 recombinant is expressed during its growth in monkey cells. The answer Goff obtained to that question is puzzling and disquieting (Table I). During the infection (late in the cycle), when both the recombinant and helper genomes are replicating, the RNA being synthesized does not hybridize to λ DNA immobilized on nitrocellulose filters, nor does it increase, appreciably, the reassociation rate of the appropriate labeled λ DNA segment. However, the same RNA does contain transcripts that hybridize with SV40 DNA; these transcripts come principally, from the helper DNA, particularly, the region replaced by the λ DNA segment.

Why is the λ DNA sequence in λ-SVGT-1 not represented amongst the RNA transcripts, when at the same time, that region of SV40 DNA in the helper is being actively transcribed? Hamer et al., (44) have also constructed a recombinant SV40 genome containing a segment of bacterial and bacteriophage DNA in approximately the same region of SV40, map position 0.74 - 1.0. In constructing their recombinant genomes, the foreign sequence was introduced by cohesive-end joining, therefore, there are no dA:dT joints bounding the inserted DNA. Monkey cells infected with these recombinants, produce RNA transcripts of the prokaryote sequences (44). This suggests that there is no inherent impediment to trans-

Synthesis of SV40- and λ-specific RNA following infection of CV-I cells with λ-SVGT-I and SV40 virus

RNA from cells infected with	% of ^{32}P-RNA hybridized to		% of total RNA homologous to	
	SV40 DNA	λ DNA	SV40 DNA	λ fragment 8 DNA
Nothing	< 0.05	< 0.02	< 0.02	< 0.0001
SV40 alone	1.97	< 0.05	0.47	< 0.0001
tsA58 plus hybrid 9	2.21	< 0.02	0.49	0.0003
tsA58 plus hybrid 18	1.20	< 0.03	0.10	< 0.0001
λ cRNA	—	35.2	—	0.86

Table 1. The first pair of columns shows the percent of the input RNA that hybridized to filters containing SV40 or λ DNA. The second pair of columns shows the fraction of the total RNA homologous to the SV40 or λ DNA as measured by the acceleration of the annealing of a labeled DNA probe by added infected cell RNA.

cription of the prokaryote DNA or gross instability of its RNA. Therefore, we have focused on two other possibilities: Conceivably, the presence, or particular arrangement of a dA:dT join (e.g. whether dA or dT is on the L-strand) blocks transcription past that site; or perhaps, RNA transcripts with a stretch of poly A or poly U near the 5' end and the complementary sequence near the 3' end, are very rapidly degraded. Experiments are currently in progress to distinguish between these and other explanations.

If dA:dT joints surrounding a DNA segment prevents expression of that sequence, the strategy for cloning using dA:dT joining will have to be reexamined. Of course DNA segments with their own transcription promoters and terminators can circumvent the effect of the dA:dT outer segments. Alternatively expression of such segments might be "turned on" by deleting either of the two dA:dT stretches.

SVGT-1 is only the first, the Mark I, of the potential SV40 transducing vectors. Various segments of the SV40 genome containing the origin of DNA replication are being tested as cloning vectors. Quite possibly, only the 100-200 base pair segment containing the origin of DNA replication could serve to propagate foreign DNA. If such an element could be replicated autonomously as a plasmid, that would eliminate the need to deal with viruses.

Before ending I want to make a few remarks about some of the opportunities and complexities of future work in this area. There is little doubt that biology has been propelled into a new era with the advent of the "recombinant DNA" methodology. The ability to join together segments of DNA from diverse sources and to propagate these recombined molecules in bacterial and animal cells will provide a deeper understanding of the structure and organization of genes in the chromosomes of all living organisms. With the molecular anatomy of the cell's genome, in hand, the perennial questions concerning chromosomal replication, activation and repression during normal cellular growth, development and differentiation can be more confidently attacked. Clarifying the molecular basis of normal and pathologic processes can make approaches to diagnosis, prevention, and cure of disease more rational and effective; opportunities for practical applications to industrial and agricultural problems are likely to follow as well. The sweeping charge made by critics of this approach, that the anticipated rewards of this research are dubious and speculative, ignores the need to acquire fresh insights in these areas of biology if we hope to ameliorate the individual and societal tragedies caused by disease.

Why then, with such promising opportunities beckoning, are we not pursuing this research as vigorously as possible? Several years ago, a group of scientists, convened by me for the U.S. National Academy of Sciences, recommended that certain recombinant DNA experimentation be deferred because potentially hazardous organisms might be produced or disseminated (45). There was, and still is, a misconception that that recommendation advocated a ban on all recombinant DNA research. That is not true. The major aim of the so-called "moratorium" was to alert scientists to a concern about the construction of bacterial strains carrying genes with information to produce cancer, certain toxins, or novel antibiotic resistance, and to secure time for a more searching analysis by individuals with greater expertise and broader concerns of any other potential risks. Unfortunately, the phrase potentially hazardous has become assuredly hazardous and might be has been changed to will be in the minds of some scientists, laymen and public officials. Not surprisingly then, the public debate has shifted from a concern of how to perform the research safely to one considering whether the research be permitted at all.

Our daily lives, societal decisions and Nation's acts carry certain risks. Similarly, the search for new knowledge cannot be risk-free. Since the answers being sought are not known in advance, the extent of the risks attendant upon discovery cannot be rigorously defined. Consequently, the assessment of risks, and the adequacy of the precautions designed to eliminate or minimize such risks, are matters of judgement based on objective, intuitive and value considerations. To curtail recombinant DNA research by relegating it to a few isolated, federally supervised, presumably impregnable, laboratories would cripple biological research and jeopardize the extraordinary opportunities that are imminent. More adamant proposals to ban recombinant DNA research throughout the world are unrealistic if not impossible to enforce. Pleas that such investigation should be postponed until the risks are known, raise questions about how data needed to assess the risks can be obtained and who will decide when all the concerns are laid to rest? Undeniably, more information is needed to assess fully the risks in this research. That information, plus the rewards, can be gained with minimal risk by expecting and requiring that scientists and institutions engaged in this work, wherever they are, adhere diligently to both the intent and letter of the guidelines governing their research.

I believe that the NIH guidelines (42) afford the security needed to meet the perceived risks. Many believe the guidelines are more stringent than what can be justified by the scientific information we now possess. The requirements imposed on scientists and their institutions by the guidelines will seriously impede progress of the research. That is a price scientists have willingly accepted; acceptance of that view is a responsible action based on careful weighing of the alternatives and rejects irrational fears as a basis for decision.

Increasingly scientists are being jarred by pressures to suppress their explorations for fear of what their discoveries will uncover or produce. The recombinant DNA controversy illustrates what a treacherous road this can be. For decisions and agreements about what is desirable, acceptable and safe to know are nearly impossible to obtain at each level of social organization. Deeply held and conflicting sociopolitical ideals challenge the traditional views of what science is for and how it should be done. As these forces gain momentum, there may be increasing attempts to restrict basic science research.

Society desperately requires effective mechanisms for anticipating and evaluating the impact of scientific and technologic breakthroughs. In the recombinant DNA matter scientists demonstrated that they could provide the early warning system for alerting society to the potential benefits and risks of their discoveries; accusations of self-interest, arrogance or even malevolence do little to encourage further efforts of that kind. Have we already squelched the concerned scientists of tomorrow? Governing bodies, everywhere, must seek better ways to encourage scientists' participation and the means to channel their input into the determination of policy.

One final note. There are so many people to thank for having made my career worth having and my life worth living. I have imposed on your time long enough so that I shall not try to acknowledge all of those debts now. But I owe my wife, Millie, a special tribute for having been so understanding and delightful a partner throughout; and an apology to my son, John, who survived his growing up years without always understanding why I was so preoccupied with other matters. And then, there is my sincere appreciation to the students, post-docs, visitors, and colleagues who shared in the adventure, and hopefully were as enriched by it as I was.

Acknowledgement

I am sincerely indebted to the National Institutes of Health, The American Cancer Society and The National Science Foundation for the generous and enlightened financial support that made my research career possible.

References

(1) Berg, P., in Reflections in Biochemistry, Ed. Kornberg, A., Horecker, B.L., Cornudella, L., and Oro, J., (Pergamon Press, New York, 1976) p. 253.

(2) Cori, C.F., ibid, p. 17.

(3) Buchanan, J. M. and Hastings, A.B. Physiol. Rev. 26 (1946) 121.

(4) Wood, H.G., ibid, 26 (1946) 198.

(5) Price, W.H., Cori, C.F., and Colowick, S.P. J. Biol. Chem. (1945) 633. Colowick, S.P., Cori, C.F. and Slein, M.W. J. Biol. Chem. 168 (1947) 583.

(6) Berg, P. and Joklik. W.K. Nature 172 (1953) 1008; J. Biol. Chem. 210 (1954) 657.

(7) Krebs, H.A. and Hems, R., Biochim. et Biophys. Acta 12 (1953) 172.

(8) Palmieri, R., Yul, R.H., Jacobs, H.K., Maland, L., Wu, L. and Kuby, S.A. J. Biol. Chem. 248 (1973) 4486.

(9) Kornberg, A. and Pricer, W.E. Jr. J. Am. Chem. Soc. 74 (1952) 1617.

(10) Jones, M.E., Lipmann, F., Hilz, H. and Lynen, F. J. Am. Chem. Soc. 75 (1953) 3285.

(11) Berg, P., J. Am. Chem Soc. (1955) 3163; J. Biol. Chem. 222 (1956) 991, 1015.

(12) Berg, P., J. Biol. Chem. 222 (1956) 1025; ibid 233 (1958) 601, 608.

(13) Berg. P., and Ofengand, E.J. Proc. Nat. Acad. Sci. U.S. 44 (1958) 78.

(14) Zamecnik, P.C. Cold Spring Harbor Symposium on Quant. Biol. 34 (1969) 1.

(15) Crick, F.H.C."On Protein Synthesis", in Symp. Soc. Exptl. Biol. No. 12 (1958) p. 138.

(16) Berg, P., Bergmann, F.H., Ofengand, E.J. and Dieckmann, M. J. Biol. Chem. 236 (1961) 1726, 1735, 1741.

(17) Chamberlin, M. and Berg, P. Proc. Nat. Acad. Sci., 48 (1962) 81.

(18) Wood, W.B., and Berg, P., Proc. Nat. Acad. Sci. 48 (1962) 94.

(19) Nirenberg, M.W., and Matthaei, J.H. Proc. Nat. Acad. Sci. 47 (1961) 1588.

(20) Cuzin, F., Vogt, M., Dieckmann, M., and Berg, P. J. Mol. Biol. 47 (1970) 317.

(21) Berg, P., The Harvey Lectures (Academic Press, New York, 1972) p. 247.

(22) Tooze, J., ed. The Molecular Biology of Tumor Viruses (Cold Spring Harbor Laboratory, Cold Spring Harbor, New York, 1973)

(23) Acheson, M.H. Cell 8 (1976) 1.

(24) Cold Spring Harbor Symposium on Quant. Biol. 39 (1974) entire volume.

(25) Nathans, D., and Smith, H.O. Ann. Rev. Biochem 44 (1975) 273.

(26) Tegtmeyer, P., and Ozer, H.L., J. Virol. 8 (1971) 516.

(27) Chou, J.Y. and Martin, R.G. J. Virol. 13 (1974) 1101.

(28) Mertz, J.E., Carbon, J., Herzberg, M., Davis, R. and Berg, P. Cold Spring Harbor Symposium Quant. Biol. 39 (1974) 69.

(29) Mertz, J.E. and Berg, P. Virology 62 (1974) 112.

(30) Lai, C.J. and Nathans, D. J. Mol. Biol. 89 (1974) 170.

(31) Carbon, J., Shenk, T.E., and Berg, P. Proc. Nat. Acad. Sci. U.S. 72 (1975) 1392.

(32) Shenk, T.E., Rhodes, C., Rigby, P.W.J. and Berg, P. Proc. Nat. Acad. Sci. U.S. 72 (1975) 989.

(33) Feunteun, J., Sompayrac, L., Fluck, M. and Benjamin, T. Proc. Nat. Acad. Sci. U.S. 73 (1976) 4169.

(34) Lai, C.J. and Nathans, D. Virology (1977) in press

(35) Van de Voorde, A., Contreras, R., Rogiers, R., and Fiers, W. Cell 9 (1976) 117.

(36) Subramanian, K.M., Dhar, R., and Weissman, S.M. J. Biol. Chem. 252 (1977) 333, 340, 355.

(37) Reed, S.I. and Alwine, J.C. Cell (1977) in press.

(38) Dhar, R., Subramanian, K.M., Pan J. and Weissman, S.M. J. Biol. Chem. 252 (1977) 368.

(39) Berg, D.E., Jackson, D.A. and Mertz, J.E. J. Virol. 14 (1974) 1063.

(40) Jackson, D.A., Symons, R.H. and Berg, P. Proc. Nat. Acad. Sci. U.S. 69 (1972) 2904.

(41) Lobban, P.E. and Kaiser, A.D. J. Mol. Biol. 78 (1973) 453.

(42) Federal Register. Part II: Department of Health, Education and Welfare, National Institutes of Health. Recombinant DNA Research Guidelines. Wednesday, July, 7, 1976.

(43) Villarreal, L., and Berg, P. Science (1977) in press.

(44) Hamer, D.H., Davoli, D., Thomas, C.A., and Fareed, G.C. J. Mol. Biol. (1976) in press.

(45) Berg, P. et al. Science 185 (1974) 303.

DNA CLONING AND PLASMID BIOLOGY

STANLEY N. COHEN, FELIPE CABELLO, MALCOLM CASADABAN,
ANNIE C.Y. CHANG, and KENNETH TIMMIS*
Department of Medicine
Stanford University School of Medicine
Stanford, California 94305 U.S.A.

Although the earliest DNA cloning experiments were reported little more than three years ago, molecular cloning of recombinant DNA has since become an important tool for the study of prokaryotic and eukaryotic biology. The initial experiments (1) involved linkage of EcoRI restriction endonuclease-generated DNA fragments to a bacterial plasmid replicon, and subsequent introduction of the composite molecule into *Escherichia coli* by transformation. Using this general method, genes from a wide variety of diverse sources have been propagated in bacteria using plasmid cloning vehicles such as pSC101 and ColE1 and their derivatives, or bacteriophage λ cloning vectors (for reviews, see refs. 2 and 3).

The advances that led to genetic manipulation of microorganisms were made in several different laboratories in the late 1960's and early 1970's. There are four basic requirements, which are illustrated in the procedure outlined in Fig. 1: 1) a replicon (cloning vehicle or vector) able to propagate itself in the recipient organism, 2) a method of joining another DNA segment to the cloning vector, 3) a procedure for introduction of the composite molecule into biologically functional recipient cells, and 4) a method of selection of bacterial cells that have acquired the hybrid DNA species.

*present address: Max Planck Institut für Molekulare Genetik, 1 Berlin 33 (Dahlem), Ihnestrasse 63-73, West Germany

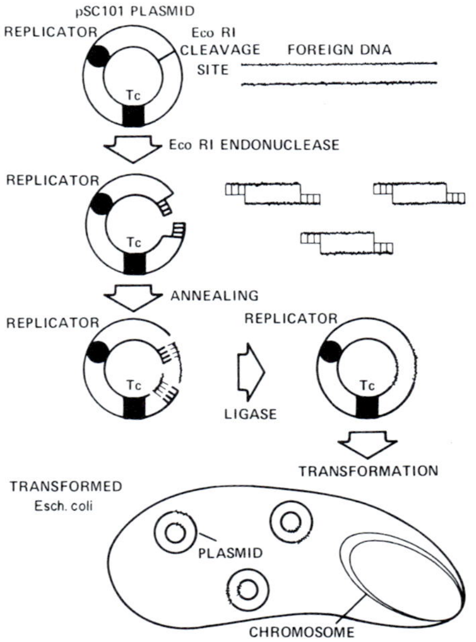

Fig. 1. DNA Cloning Procedure. The pSC101 plasmid, which carries replication functions, a gene for tetracycline resistance, and a single cleavage site for the EcoRI restriction endonuclease is cut by the enzyme at a site that does not interfere with replication functions or tetracycline resistance. Cleavage results in the formation of short complementary single strand DNA termini. Similarly-cleaved foreign DNA is annealed with pSC101 plasmid DNA, and nicks are sealed by the enzyme DNA ligase. The resulting molecule is transformed into calcium chloride-treated *E. coli*, and selection is carried out for Tc-resistant cells. The cloning vehicle and the inserted foreign DNA fragments are propagated in the progeny of the original transformed cell.

At the present time, a wide variety of experimental alternatives are available for cloning of foreign DNA segments in E. coli. The essential requirement in all the procedures is a cloning vector replicon able to propagate itself and an attached segment of foreign DNA in the recipient organism; thus, the cloning vector must contain an appropriate replication "origin" and associated genetic functions. Various natural and constructed plasmids and appropriate mutants of bacteriophage lambda have been used for this purpose. Most conveniently, the construction of biologically functional recombinant plasmids can be carried out using cohesive ends (4,5) generated by a restriction endonuclease that cleaves the plasmid at a single site; some plasmids commonly used as vectors are cleaved once by each of several endonucleases, making these plasmids useful in the cloning of DNA fragments generated by different enzymes. In some instances, the cleavage site is located in a gene that is phenotypically detectable (eg. ColE1 (6), tetracycline (7,8,9), kanamycin, (Timmis, Cabello and Cohen, manuscript in preparation)). This facilitates the detection of clones carrying foreign DNA fragments through insertional inactivation of a gene on the vector (10).

Phage lambda mutants suitable for the cloning of either HindIII or EcoRI-generated DNA fragments have been isolated (11,12,13); with certain mutants, the formation of viable phage particles is dependent on insertion of a foreign DNA segment, facilitating the detection of chimeric DNA molecules. Although suitable lambda mutants are currently not available for use with other restriction endonucleases, the use of adapter fragments (9), as described below, potentially circumvents this problem and allows considerable flexibility.

Biologically functional DNA molecules can also be constructed by addition of a series of identical deoxyribonucleotides (eg. dA) to the ends of one DNA species and addition of complementary deoxyribonucleotides (eg. dT) to the ends of the second species. Use of the dA-T "terminal transferase" procedure (14,15) prevents the joining together of separate fragments derived from the same DNA molecule, facilitating the joining of such fragments to the cloning vector (16). While fragments of DNA joined by the dA-T method cannot be liberated at a unique cleavage site, as is the case when restriction endonuclease-generated cohesive ends are used for linkage (1), a recent report indicating that the Aspergillus S1 single-stranded endonuclease cleaves phage lambda DNA preferentially in A-T-rich regions under conditions of appropriate temperature and ionic strength (17), suggests that foreign DNA fragments inserted into cloning vectors by the d-AT joining method potentially can also be enzymatically excised.

A method that involves the addition of the dC-G termini of DNA molecules lacking cohesive ends (18), plus utilization of the short cohesive ends generated by the *Eco*RI restriction endonuclease under conditions of low salt potentially enables the reconstruction of *Eco*RI-cleavable termini on cloned fragments of DNA. Variations of this method seem likely to enable coupling of dA-T termini to cohesive ends generated by certain other restriction enzymes — with resulting reconstruction of endonuclease cleavage sites at the junction. Thus, there are a wide variety of ways of making the "joints" required for molecular cloning of recombinant DNA. However, while cohesive termini are convenient for some experiments, they are not required for the linkage of DNA fragments; blunt ended molecules can be joined using the T4 DNA ligase (19). The discovery of blunt-ended joining (19), together with the demonstration that a multitude of different restriction endonucleases cleave DNA at various sites to yield blunt-ended fragments, provides enormous flexibility in the *in vitro* construction of recombinant DNA molecules (9,20,21,22).

While ligation *in vitro* appears to increase the efficiency of joining of DNA fragments to the cloning vector, *in vitro* ligation may not be required when selectable phenotypic traits are present on the fragments to be cloned, at least for cohesive-ended fragments; *in vivo* ligation of separate *Eco*RI endonuclease-generated fragments of the R6-5 plasmid has been shown following transformation (1). It remains to be determined whether *in vivo* ligation of non-cohesive-ended fragments can occur.

Chimeric plasmids can be introduced into recipient cells by means of a plasmid DNA transformation procedure (20); plasmid transformation has enabled the cloning of individual genomes, making it possible to study the progeny of single DNA molecules in ways that previously were practical only with infective particles such as viruses. With this procedure, the progeny of individual molecules of plasmid DNA can be amplified many fold, and can be propagated indefinitely in bacterial cultures. The procedure commonly employed depends on the ability of divalent cations such as calcium chloride (24) or barium chloride (25) to alter membrane permeability of bacteria — thereby enabling them to take up plasmid DNA molecules (23). When the entering plasmid carries a replication system capable of functioning in the recipient cell, the plasmid can propagate itself and other DNA fragments that are attached to it. Bacteriophage vectors are also taken up by appropriately treated cells by a process called _transfection_, but in this case cloning of inserted DNA fragments does not require survival of recipient cells — such cells need only yield viable viral particles consisting of the phage replicon plus the foreign DNA segment. Introduction of phage chimeras into bacterial cells potentially can be accomplished using one of several transfection procedures that employ calcium chloride treatment (23), spheroplast production (26), or a "helper phage" assay (27).

Separate antibiotic resistance genes carried by both the plasmid cloning vehicle and by the inserted DNA fragments were initially used to select for bacterial cells that had acquired the chimeric molecule (1). Other phenotypic markers such as colicin immunity or metabolic characteristics have since been used (6,16,28). As noted above, certain plasmid genes contain restriction endonuclease cleavage sites within them, and inactivation of such genes by insertion of a foreign DNA fragment can be employed to select for chimeric plasmids (10). Certain plasmid vectors have been designed especially to utilize insertional inactivation; the pAC181 plasmid (Figure 2) is one of these. pAC181 was formed by linking the replication functions of the mini-circular plasmid from _E. coli_ strain 15 to the chloramphenicol resistance gene of R6-5 and the tetracycline resistance gene of pSC101. The resulting plasmid has an _Eco_RI cleavage site within the chloramphenicol resistance gene, and insertion of DNA fragments within _Hind_III, _Bam_I, or _Sal_I cleavage sites of the plasmid inactivate the tetracycline resistance gene. Inactivation of the resistance markers is easily detected by plating procedures, or selected using modifications of a penicillin-selection procedure (29). In addition to being suitable for use with the insertional inactivation procedure for _Eco_RI, _Hind_III, _Bam_I, or _Sal_I DNA fragments, the pAC181 plasmid is multicopy, is amplifiable, and is non-conjugative (Chang and Cohen, manuscript in preparation).

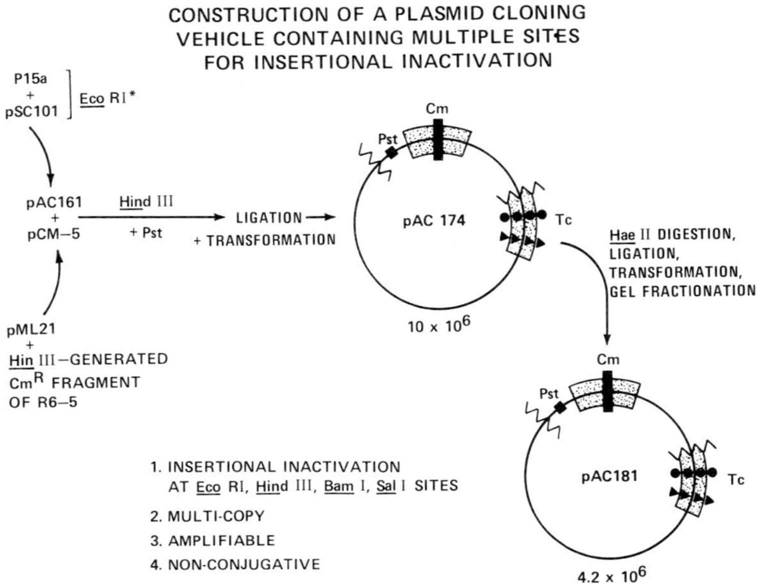

Fig. 2. Construction of pAC181 Cloning Vehicle. Construction of the plasmid is described in the text. This cloning vehicle is suitable for use with the insertional activation procedure for EcoRI, HindIII, BamI, or SalI DNA fragments. In addition, it is a non-conjugative, multicopy cloning vector that is amplifiable in the presence of appropriate concentrations of chloramphenicol.

A wide variety of other methods have been used to enable selection of plasmids carrying specific DNA fragments; in general, these involve the use of radioactively labelled DNA or RNA probes or immunological probes to identify clones carrying the desired chimeric plasmids or phage. The subculture-cloning selection procedure employed for the cloning of histone genes from unfractionated sea urchin DNA (30) is essentially a modification of the sib-selection technique reported some years ago by Cavalli-Sforza and Lederberg (31) for the genetic identification of bacterial mutants.

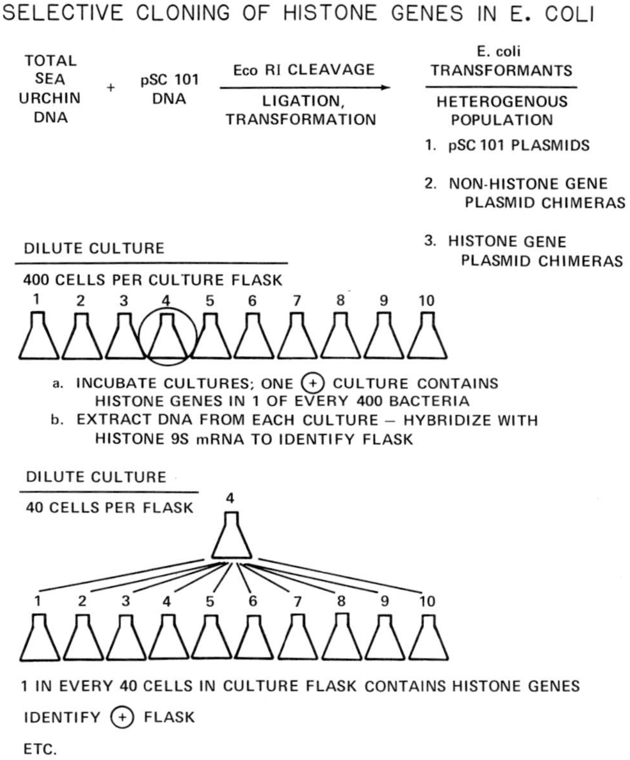

Fig. 3. Use of Sub-Culture Cloning Procedure for Isolation of Desired Gene from Unfractionated DNA Using Radioactively-Labelled Probe. The procedure was used by Kedes et al. (30) for purification of sea urchin histone genes. The method is described in the text.

Figure 3 diagrams the principle involved in the use of the procedure for the identification of chimeric plasmids carrying sea urchin histone genes. An unfractionated population of restriction endonuclease-generated DNA fragments was ligated to the pSC101 plasmid vector, transformation of *E. coli* was carried out, and a heterogeneous population of *E. coli* transformants that expressed the tetracycline resistance gene of the vector was selected by addition of this antibiotic to cultures. In these circumstances, certain of the bacteria would be expected to carry only the recircularized pSC101 vector; others will carry chimeric plasmids consisting of the vector plus a non-histone gene fragment, and the experimental task is to identify these; this is accomplished by use of radioactively-labelled messenger RNA probes specific for histone gene DNA. The heterogeneous population is divided into a series of sub-cultures as shown in the diagram, and any sub-culture containing bacterial cells that carry the desired gene(s) are identified by DNA-RNA hybridization. Because a limited number of cells were initially placed in each sub-culture, this procedure results in enrichment of the population for bacteria containing the desired chimeric plasmids. Several cycles of the procedure are sufficient to allow cloning of the desired DNA fragment.

Once the population of cells has been partially enriched for the chimeric plasmid by sub-culture cloning, the specific cells containing the desired molecule can be identified using in situ hybridization procedures (32). These are also applicable for the identification of bacteriophage plaques formed by viruses that carry inserted DNA fragments (33). Recently, immunological procedures for identifying plaques made by phage carrying specific inserted fragments have also been described (34).

Recently, the flexibility of DNA cloning procedures has been increased considerably by the use of "adapter fragments" containing termini generated by two different restriction endonucleases. One source of such adapter fragments is the pSC101 plasmid, which contains a series of cleavage sites for different restriction endonucleases within close proximity to each other (7,8,9). Because this plasmid includes single EcoRI, HindIII, BamI, and SalI cleavage sites within a region about 800 nucleotides in length, adapter fragments can be produced by treatment of pSC101 DNA with different combinations of the enzymes (Figure 4).

USE OF ADAPTER FRAGMENTS IN DNA CLONING

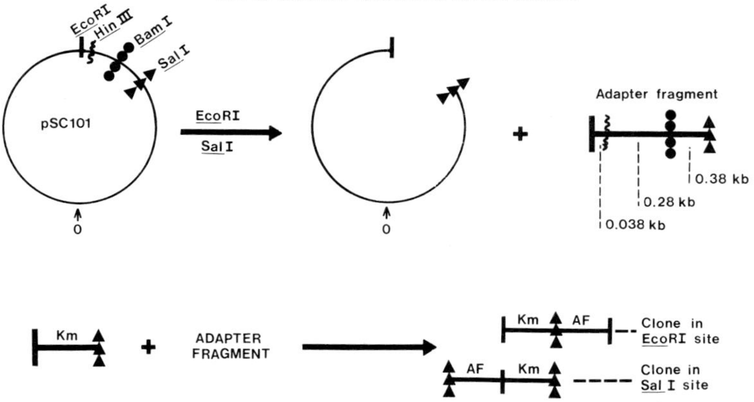

USE OF ADAPTER FRAGMENTS FOR SUBSTITUTION OF RESTRICTION ENDONUCLEASE SITES

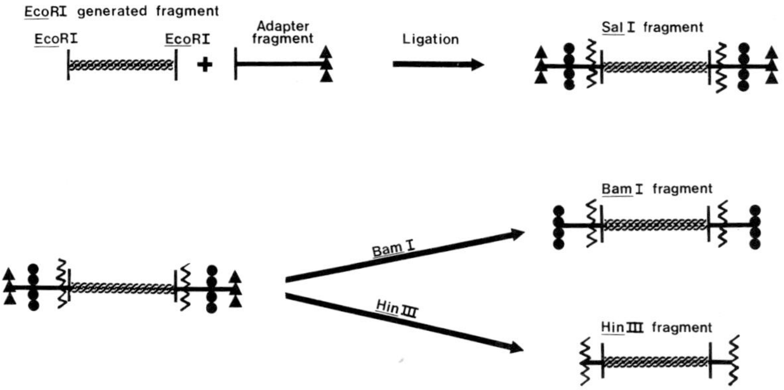

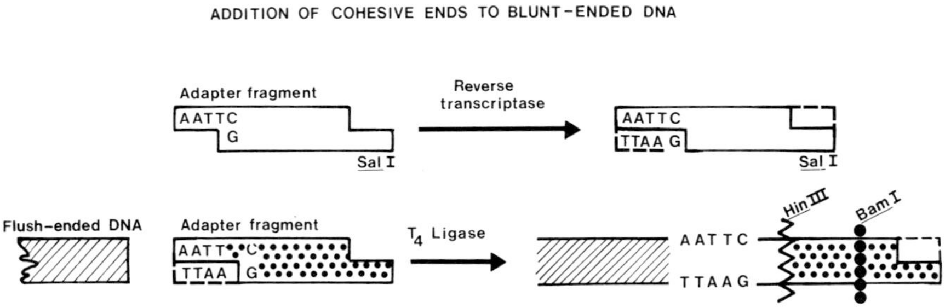

Fig. 4. Use of Adapter Fragments in DNA Cloning. The pSC101 plasmid contains EcoRI, HindIII, BamI and SalI restriction endonuclease cleavage sites in close proximity to each other. The location of these sites is shown in relation to the pSC101 origin of replication (0). Treatment of pSC101 DNA with both the EcoRI and SalI enzymes leads to formation of the structures shown. The adapter fragments resulting from cleavage by the two enzymes can be used to convert a terminus produced by one restriction endonuclease to an end that can be cloned at a cleavage site generated by another enzyme. Use of adapter fragments for substitution of restriction endonuclease cleavage sites and for the addition of cohesive termini blunt-ended DNA fragments is illustrated.

Such adapter fragments are potentially useful for conversion of the cleavage site generated by one of the restriction enzymes into a terminus that can be joined to a fragment produced by digestion of DNA with another enzyme. The utility of adapter fragments in DNA cloning experiments is further enhanced by the observation that blunt-ended DNA molecules can be joined together by the bacteriophage T4 ligase (19), as noted above. The use of blunt-ended ligation and adapter fragments from the pSC101 plasmid enables the insertion of blunt-ended DNA fragments into sites cleaved by the EcoRI, HindIII, BamI, or SalI restriction endonucleases. These procedures have been used for restriction of the pAC181 plasmid shown in Figure 2. They are also applicable for the cloning of cDNA segments or short polynucleotide fragments synthesized *de novo*.

USE OF DNA CLONING FOR STUDY OF PLASMID BIOLOGY

DNA cloning procedures have been widely used for the introduction of DNA from a wide variety of eukaryotic organisms into *E. coli*, and the experiments carried out have provided considerable information about the organization of genetic information on the chromosomes of complex cells. As noted above, recombinant DNA procedures have also been used for the construction of plasmid vectors designed to fulfill specific experimental needs, such as insertional inactivation or reduced transmissability. In addition, DNA cloning has proven to be a useful technique for the study of the biology of plasmids themselves, and this has been especially true in investigations of DNA replication.

REPLICATION STUDIES

Although naturally occurring plasmids carry a wide variety of genes ranging from metabolic degradative pathways to antibiotic resistance traits (35), a feature common to all plasmids is a system that accomplishes autonomous replication of the plasmid as an extrachromosomal element. Plasmids can thus be considered as replication systems (that is, a replication origin and associated genetic functions) that are linked to other genes conducive to survival of the plasmid. Certain large plasmids are known to carry more than one origin of replication. However, little has been known about the functional interaction of multiple replication systems that may coexist on a single DNA molecule.

DNA cloning procedures have enabled the use of two novel approaches to the study of plasmid DNA replication: the first has been the construction of model replicons in which two physically and functionally distinct and well-characterized replication systems have been fused by joining them at restriction endonuclease cleavage sites. The second is a "paring-down" of complex replicons by the specific selection of those segments of the plasmid that carry replication functions. Figure 5 diagrams both of these approaches.

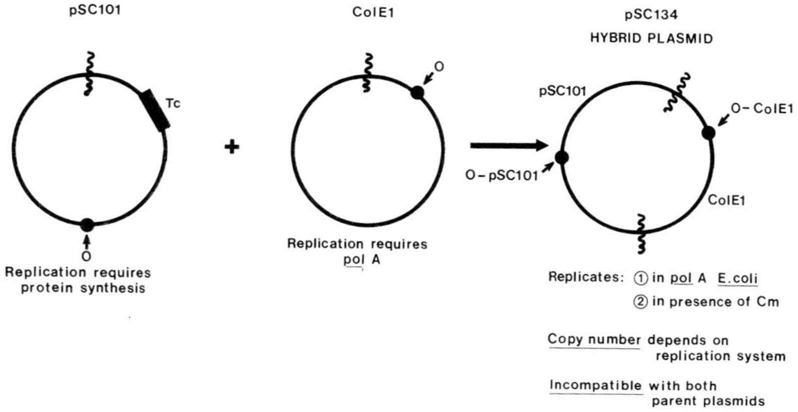

ISOLATION OF REPLICATION REGIONS OF COMPLEX GENOMES

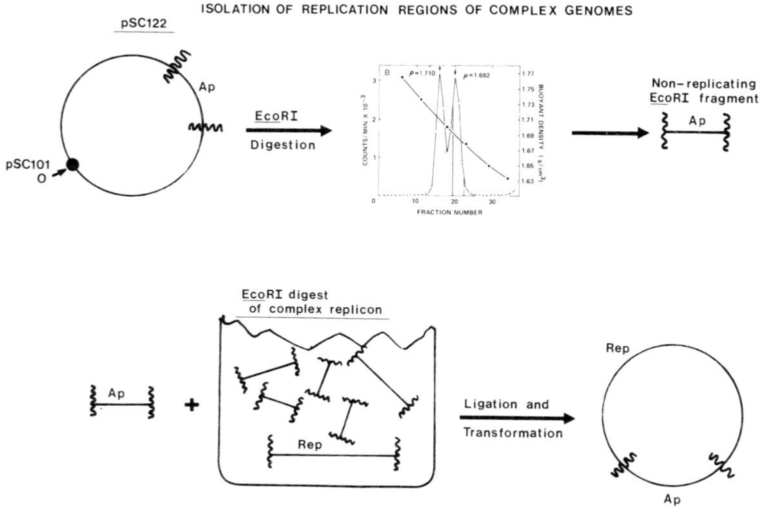

Fig. 5. Utilization of DNA Cloning Procedures in the Study of Plasmid Replication. The figure diagrams the method used for construction of composite plasmids containing two functional replication systems and the procedure used for isolation of replication region fragments of complex genomes.

Using the pSC134 composite plasmid, which was constructed by *in vitro* linkage of the colicin-producing plasmid ColE1 and the tetracycline-resistance plasmid pSC101 at their unique *Eco*RI restriction endonuclease cleavage sites (10), it was possible to show that each component replication system of the hybrid plasmid could accomplish replication of the entire molecule. However, studies of replication origin and usage by the pSC134 plasmid indicated that under normal conditions it utilizes only its ColE1 replication system. Under such conditions, the composite plasmid is nevertheless *incompatible* with both of the parent replicons. Such studies, together with studies of copy-number control in the composite plasmid, support the view that a *trans*-dominant gene product is involved in the determination of plasmid incompatibility.

The experimental plan for such studies of the isolation of plasmid fragments that carry the replication functions of the molecule is shown in Figure 5. The procedure involves the use of a selectable, but non-replicating endonuclease-generated DNA fragment as a probe to enable the isolation and identification of DNA segments capable of autonomous replication; in our investigations (36) a fragment of staphylococcal plasmid DNA carrying a gene for resistance to ampicillin (Ap) and penicillin (Pc) (37) was used as a probe. Although this fragment was derived from another bacterial species, it codes for an antibiotic resistance trait indigenous to *E. coli*, and this trait is expressed in the *E. coli* host. In other investigations (38) a Km-resistance fragment originally derived from the R6-5 plasmid (1) has been used as a probe for the isolation of plasmid replication regions. Using those procedures, it has been possible to isolate *Eco*RI-generated DNA fragments carrying the replication regions of the R6-5 and F*lac* plasmids (36,38). Although the replication region fragments contain only about 10% of the nucleotide sequences of the parent genomes, they were shown to specify both incompatability and copy number control properties of the parent plasmids as well as other replication-related plasmid properties (36).

The general methods employed for isolation and characterization of replication regions of the R6-5 and F*lac* plasmids are potentially applicable for the isolation of DNA segments containing the replication origin and/or genes of any complex replicon capable of functioning in microorganisms, and they may also be useful in the study of chromosomal replication. This sequestration of replication functions of large genomes onto small plasmid DNA molecules should potentially facilitate *in vitro* and *in vivo* investigations of the gene products involved in DNA replications. With appropriate modification, the principle of using a restriction endonuclease-generated "probe" fragment that lacks a particular genetic function to select for another fragment carrying the function may also permit identification and isolation of other phenotypically defined regions of complex genomes.

ISOLATION OF PROMOTER REGIONS OF PLASMIDS

Certain restriction endonuclease cleavage sites on plasmids are known to separate promoter regions from the function genes associated with the promoter. In particular, the *Hind*III cleavage site adjacent to the structural gene for tetracycline resistance in the pSC101 plasmid and the *Hind*III cleavage site in the kanamycin operon of pSC105 and its derivatives appear to be located between promoters from structural genes (Chang and Cohen, unpublished data; H. Boyer, personal communication; A. Rambach, personal communication; ref. 22). In certain instances, insertion of a DNA fragment into such cleavage sites blocks expression of kanamycin and/or tetracycline resistance --- while insertion of other fragments permits expression to occur. Presumably, the ability of a DNA fragment to permit readthrough from a promoter on one side of the cleavage site to a structural gene on the other side depends on whether termination signals and/or a functional promoter region exist within the inserted fragment.

Recently, it has been possible to design and construct a hybrid plasmid that is particularly suitable for the detection of inserted DNA fragments carrying signals for the initiation and/or termination of RNA synthesis. These experiments combine the *in vivo* gene fusion techniques described by Casadaban (39) with *in vitro* recombinant DNA procedures. Specifically, the constructed plasmid (pMC81, Fig. 6) contains: 1) the *lac* genes of *E. coli* under control of the *ara* operon promoter, and 2) an intervening DNA fragment containing a *Hind*III cleavage site.

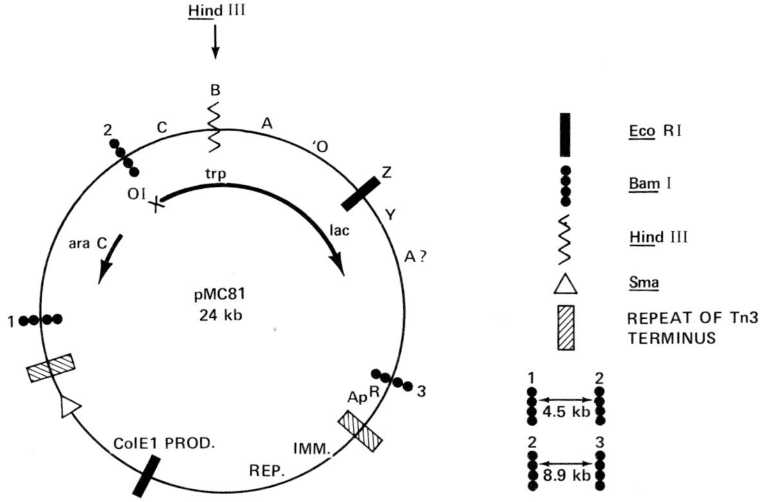

Fig. 6. Diagram of pMC81 Plasmid Constructed for the Detection of DNA Segments Containing Signals for the Initiation or Termination of RNA Synthesis. A HindIII cleavage site separates the lac genes from the arabinose operon promoter. Insertion of a DNA fragment having a termination signal at this site prevents arabinose-induced expression of lac. Insertion of a fragment carrying a promoter region results in lac expression in the absence of arabinose induction.

In this system, expression of lac requires induction with arabinose. Insertion of a fragment having a termination signal at the intervening HindIII cleavage site prevents arabinose-induced expression of lac. Insertion of a fragment carrying a promoter region results in lac expression in the absence of arabinose induction. The use of reverse transcriptase and blunt-ended ligation procedures potentially allows the HindIII cleavage site in the pMC81 plasmid to be used with DNA fragments generated by a wide variety of restriction endonucleases. In experiments carried out thus far, the pMC81 plasmid has been employed for the isolation and study of promoter regions on complex plasmid replicons such as R6-5.

INSTABILITY OF CERTAIN CHIMERIC PLASMIDS

During studies of chimeric plasmids containing various kinds of eukaryotic or prokaryotic DNA, we observed that some chimeric plasmids were lost from bacterial cells in the absence of continued selection for the antibiotic resistance gene carried by the cloning vehicle segment of the plasmid. Moreover, even when continued selection for a gene on the cloning vehicle was carried out, the inserted foreign gene segment appeared to have been lost in at least some cases.

These phonomena are illustrated in Figure 7, which shows the relative stability of these different composite plasmids constructed entirely of prokaryotic DNA segments: bacteria were grown in the absence of antibiotics for the number of generations shown, and were then tested for expression of antibiotic resistance determinants as indicated. The pSC134 plasmid, which is a hybrid DNA molecule constructed by linkage of the pSC101 and ColE1 replicons, was stable in the absence of selection --- as was the pSC122 plasmid, which contains an ampicillin-penicillin resistance fragment of staphylococcal DNA inserted into pSC101. However, another plasmid (pSC112), which contains a different staphylococcal Ap-resistance fragment (37), appears to be <u>unstable</u>; growth of bacteria containing pSC112 in the absence of antibiotics led to a gradual loss of the Tc-resistance specified by the pSC101 cloning vehicle and a much more rapid loss of the Ap-resistance trait carried by the inserted DNA fragments. The observation that Tc-resistant, Ap-sensitive cells were produced in cultures of bacteria carrying a plasmid that was shown by heteroduplex analysis and agarose gel electrophoresis to originally contain the staphylococcal DNA Ap-resistance fragment (37) suggested that loss of the staphylococcal Ap-resistance gene component of the chimeric plasmid night be occurring in some cells. Isolation of characterization of DNA from Tc-resistant, Ap-sensitive cells (Cabello, Timmis, and Cohen, in preparation) has confirmed this interpretation.

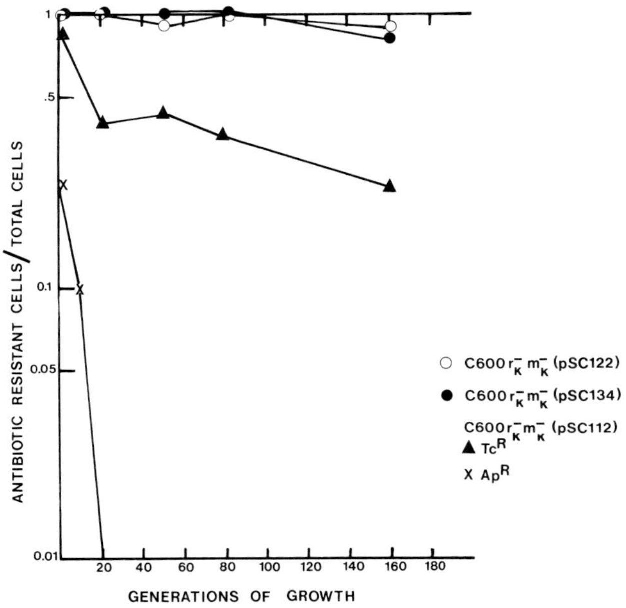

Fig. 7. Instability of Certain Chimeric Plasmids in the Absence of Selective Pressure. Cultures of E. coli strain C600 rk^-mk^- carrying the E. coli hybrid plasmid pSC134 or the E. coli-S. aureus plasmids pSC112 or pSC122 were grown in media containing antibiotics, and were then diluted at time 0 into culture media lacking drugs. Serial transfer was carried out for more than 100 generations of growth in the absence of antibiotics. Samples were removed as indicated, and the fraction of cells expressing the Tc-resistance of the pSC101 component of each plasmid, and the Ap-resistance inserted staphylococcal plasmid DNA fragment was determined.

It has been previously reported (40) that mouse mitochondrial DNA-pSC101 hybrid plasmids could undergo a rearrangement of plasmid DNA under certain experimental conditions. Recombination, DNA inversion, and transposition (translocation) of DNA segments also appeared to occur within other chimeric plasmids even in recA$^-$ bacteria (Chang and Cohen, unpublished data). The reassortment/rearrangement of DNA sequences may occur more frequently with chimeric plasmids, but it also seem to occur in naturally occurring *E. coli* plasmids. During some of the earliest cloning experiments, an EcoRI-generated fragment of R6-5 carrying resistance to sulfonamide, but not to streptomycin, was cloned. In later experiments, in which the EcoRI-generated fragments of the R6-5 plasmid were each individually cloned into a ColE1-derived vector (Timmis, Cabello and Cohen, in preparation), an EcoRI-generated sulfonamide-resistance fragment was isolated --- but this fragment additionally carried a resistance gene to streptomycin. In other studies, (Timmis, Cabello, and Cohen, in preparation) variability in the length of corresponding fragments cloned from different preparations of the R6-5 plasmid has been noted, and even a difference in the restriction endonuclease cleavage pattern has been observed with different preparations of this plasmid. These findings are analogous to the loss or gain of the restriction endonuclease sites observed in prokaryotic-eukaryotic chimeric plasmids, as noted above.

Several years ago, at the time of construction of the first interspecies chimeric plasmids (37), it was observed that the pSC112 plasmid, which was formed by insertion of an Ap-resistance segment of the staphylococcal pI258 plasmid into the EcoRI cleavage site of pSC101, expressed tetracycline resistance at the same level as the pSC101 parent. However, an analogous plasmid (pSC113) which had two other segments of pI258 inserted into the EcoRI cleavage site of pSC101, expressed Tc resistance at a very much lower level. A tetracycline resistance plasmid having physical and biological properties (including Tc-resistance levels) indistinguishable from those of pSC101 could be recovered after transformation of *E. coli* by an EcoRI digested pSC113 DNA. Although it had been shown earlier that inclusion of a DNA fragment into the single EcoRI cleavage site of pSC101 did not interrupt the Tc-resistance gene of the plasmid (1), it appeared from these experiments that in at least some instances insertion of a fragment at this site would affect the level of expression of Tc-resistance (37, 41). Subsequent experiments (Tait et al., submitted for publication; Cabello et al., manuscript in preparation) have provided more direct evidence that insertion of DNA into the EcoRI cleavage site of pSC101 influences the level of expression of tetracycline-resistance by the plasmid. Whereas Tc-resistance of pSC101 is inducible, insertion of certain DNA fragments into this site prevents inducibility and insertion of other fragments results in constitutive expression of high level tetracycline resistance.

ACKNOWLEDGEMENTS

These studies were supported by grants AI08619 from the NAID, BMS 75-14176 from the NSF and VC139A from the American Cancer Society to S.N.C. K.T. was a postdoctoral fellow of the Helen Hay Whitney Foundation and M.C. is a Dernham Junior Fellow of the California Division of the American Cancer Society.

REFERENCES

(1) S.N. Cohen, A.C.Y. Chang, H.W. Boyer and R.B. Helling, Proc. Natl. Acad. Sci., U.S.A., 70 (1973) 3240-3244.

(2) R. Curtiss,III, in: Annual Review of Microbiology, Vol. 30 (Annual Reviews, Inc., Palo Alto, 1976) p. 507-533.

(3) R.L. Sinsheimer, Annual Review of Biochemisty, (1977) (in press).

(4) J.E. Mertz and R.W. Davis, Proc. Natl. Acad. Sci., U.S.A., 69 (1972) 3389-3393.

(5) V. Sgaramella, Proc. Natl. Acad. Sci., U.S.A., 69 (1972) 3370-3374.

(6) V. Hershfield, H.W. Boyer, C.Yanofsky, M.A. Lovett and D.R. Helinski, Proc. Natl. Acad. Sci., U.S.A., 71 (1974) 3455-3459.

(7) D.H. Hamer and C.A. Thomas, Jr., Proc. Natl. Acad. Sci., 73 (1976) 1537-1541.

(8) R.L. Rodriguez, F. Bolivar, H.M. Goodman, H.W. Boyer and M. Betlach, ICN-UCLA Symposium on Genetic Regulatory Mechanisms (Academic Press, New York, 1976) (in press).

(9) S.N. Cohen, F. Cabello, A.C.Y. Chang and K. Timmis, Proceedings of the Miles Symposium on Recombinant DNA (1977).

(10) K. Timmis, F. Cabello and S.N. Cohen, Proc. Natl. Acad. Sci., U.S.A., 71 (1974) 4556-4560.

(11) A. Rambach and P. Tiollais, Proc. Natl. Acad. Sci., U.S.A., 71 (1974) 3927-3930.

(12) N.E. Murray and K. Murray, Nature, 251 (1974) 476-481.

(13) M. Thomas, J.R. Cameron and R.W. Davis, Proc. Natl. Acad. Sci., U.S.A., 71 (1974) 4579-4583.

(14) P.E. Lobban and A.D. Kaiser, J. Mol. Biol., 78 (1973) 453-471.

(15) D.A. Jackson, R.H. Symons and P. Berg, Proc. Natl. Acad. Sci., U.S.A., 69 (1972) 2904-2909.

(16) L. Clarke and J. Carbon, Proc. Natl. Acad. Sci., U.S.A., 72 (1975) 4361-4365.

(17) V.N. Ksënzenko, L.P. Tikhomirova and N.I. Matvienko, Nucl. Acids Res., 3 (1976) 1081-1093.

(18) F. Rugeon, P. Kourilsky and B. Mack, Nucl. Acids Res., 2 (1975) 2365-2378.

(19) V. Sgarmella and H.G. Khorana, J. Mol. Biol., 72 (1972) 427-444.

(20) K.J. Marians, R. Wu, J. Stawinski, T. Hozumi and S.A. Narang, Nature, 263 (1976) 744-748.

(21) H.L. Heyneker, J. Shine, H.M. Goodman, H.W. Boyer, J. Rosenberg, R.E. Dickarson, S.A. Narang, K. Itakura, S. Lin and A.D. Riggs, Nature, 263 (1976) 748-752.

(22) K. Backman, M. Ptashne and W. Gilbert, Proc. Natl. Acad. Sci., U.S.A., 73 (1976) 4174-4178.

(23) S.N. Cohen, A.C.Y. Chang and L. Hsu, Proc. Natl. Acad. Sci., U.S.A., 69 (1972) 2110-2114.

(24) M. Mandel and A. Higa, J. Mol. Biol., 53 (1970) 159-162.

(25) A. Taketo, Z. Natürforsch, 30C (1975) 520-522.

(26) G.D. Guthrie and R.L. Sinsheimer, J. Mol. Biol, 2 (1960) 297-305.

(27) A.D. Kaiser and D.S. Hogness, J. Mol. Biol., 20 (1966) 183-209.

(28) B. Polisky, R.J. Bishop and D.H. Gelfand, Proc. Natl. Acad. Sci., U.S.A., 73 (1976) 3900-3904.

(29) B.D. Davis, American Chemical Society J., 70 (1948) 4367.

J. Lederberg and N. Zinder, American Chemical Society J., 70 (1948) 4367-4368.

(30) L.H. Kedes, A.C.Y. Chang, D. Housman and S.N. Cohen, Nature, 255 (1975) 533-538.

(31) L.L. Cavalli-Sforza and J. Lederberg, Genetics, 41 (1956) 367- 381.

(32) M. Grunstein and D.S. Hogness, Proc. Natl. Acad. Sci., U.S.A., 72 (1975) 3961-3965.

(33) K. Jones and K. Murray, J. Mol. Biol., 96 (1975) 455-460.

(34) B. Sanzey, O. Mercereau, T. Ternynck,and P. Kourilsky, Proc. Natl Acad. Sci., U.S.A., 73 (1976) 3394-3397.

(35) S.N. Cohen, Nature, 263 (1976) 731-738.

(36) K. Timmis, F. Cabello and S.N. Cohen, Proc. Natl. Acad. Sci., U.S.A., 72 (1975) 2242-2246.

(37) A.C.Y. Chang and S.N. Cohen, Proc. Natl Acad. Sci., U.S.A., 71 (1974) 1030-1034.

(38) M.A. Lovett and D.R. Helinski, J. Bacteriol., 27 (1976) 982-987.

(39) M. Casadaban, J. Mol. Biol., 104 (1976) 541-555.

M. Casadaban, J. Mol. Biol., 104 (1976) 557-566.

(40) A.C.Y. Chang, R.A. Lansman, D.A. Clayton and S.N. Cohen, Cell, 6 (1975) 231-244.

(41) R.C. Tait, R.L. Rodriguez and H.W. Boyer, submitted for publication.

DISCUSSION

L. KODITSCHECK: I believe you said that in some cases there was more than one replicating site in a plasmid, is that right? Does that have any relation to the number of copies of the plasmid per cell?

S.N. COHEN: No, it does not appear to. However, the pSC134 plasmid is a composite of pSC101 plus ColE1. Ken Timmis, and Felipe Cabello in my laboratory show that this plasmid ordinarily used only its ColE1 replication system, and its copy number is maintained at the copy number of ColE1. If one turns off the ColE1 replication system by, for example, putting the pSC134 plasmid in a bacterial host that makes a temperature sensitive DNA polymerase I and raising the temperature, the pSC101 replication system is turned on and the copy number of the composite plasmid is maintained at the copy number of pSC101. Thus, although this composite plasmid is capable of using both of its origins of replication, it ordinarily uses only one. However, while using only the ColE1 replication system the plasmid is incompatible with both of its parents, suggesting involvement of a diffusible trans acting substance in determining incompatibility.

R. FAUST: I have recently returned from the U.S.S.R. where my Russian counterparts and I have drafted a comprehensive research program dealing with my on-going recombinant studies of bacterial insect pathogens in attempts to develop broader spectrum and safe biological insecticides for commercial production. This program is a part of the joint US-USSR Science and Technology Agreement supported by the National Science Foundation and the Armenian SSR Academy of Sciences. During our deliberations the problem of maintaining viability of exchange materials was discussed. Have you knowledge of the stability of recombinant plasmids in a freeze-dried state and what is your method of preservation of plasmid materials prior to experimental use; if any?

S.N. COHEN: No, we have not studied that. However, in regard to this point, it should be mentioned that there was a recent note in *Science* proposing the creation of a chimeric plasmid gene bank and soliciting reactions to the proposal. I would guess that the stability of any chimeric plasmid in such a bank would depend on the method of storage and the particular plasmids being considered. In our experience, the stability seems to vary considerably from plasmid of plasmid. Certain plasmids aren't stable while others appear to be stable. For example, the two plasmids consisting of sea urchin histone genes from *S. purpuratus* and pSC101 appear to be quite stable, but one containing *L. pictus* histone genes inserted at the same *Eco*RI cleavage site of pSC101 seems to undergo recombinational events. The basis for these differences isn't clear to us.

C. WEISSMANN: Are there any data on the actual preservation of the sequences inserted over many generations of replication?

S.N. COHEN: We have not done DNA sequence analysis of cloned fragments of DNA that have been propagated in *E. coli* for varying periods of time. However, at the level of resolution permitted by heteroduplex analysis and restriction enzyme cleavage patterns, some of these DNA species appear to be quite stable. In some instances, the stability depends on which host is used, and the function of the *E.coli* *rec*A gene does not appear to play a significant part, as was observed in the studies of mouse mitochondrial DNA that Annie Chang, Bob Lansman, Dave Clayton and I have reported. Brown *et al*. have observed that mouse mitochondria DNA inserted into the pSC101 plasmid appeared to be quite stable under the conditions they used. I don't want to overemphasize this instability of recombinant DNA plasmids and rearrangement of DNA sequence, but I do want to say that we have seen a number of instances where there has been loss of restriction sites or addition of new restriction sites. Because of the *rec*A-independence of these events, we think that so-called "illegitimate" recombinational processes are involved. In some instances, as I showed in the last photomicrograph presented in my talk, inverted repeat DNA sequences appear to be showing up in composite plasmids.

C. WEISSMANN: Some variations may only be picked up when you start recloning after many generations. Because of the mass population, you may not notice their appearance.

S.N. COHEN: That's right - except that the composite plasmid may be at a disadvantage in comparison with a plasmid lacking the inserted fragment. In that case, the loss of fragment is going to result in a molecule that can propagate itself beter than the composite, and it will be the predominant molecule species after a number of generations. Perhaps, this process is going on naturally in plasmids all the time, but most of the time the loss of a fragment does not yield a molecule that has an advantage over the original molecule. One would expect the process of evolution to have gotten rid of disadvantageous fragments in natural plasmids long ago, so that plasmids isolated from natural sources will probably be fairly stable. However, when a new plasmid molecule is constructed by recombinant DNA techniques,we might expect that it has to go through an evolutionary process that excises undersirable fragments of DNA until it reaches a stable form.

F. BLATTNER: Have you succeeded in transplanting the transposable elements into anything other than prokaryotic DNA?

S.N. COHEN: We find that the transposable element goes into eukaryotic sequences that have been cloned on bacterial plasmids, so that there is prokaryotic-eukaryotic recombination occuring by this mechanism within bacterial cells.

A. EISENSTARK: You mentioned that the illegitimate recombination event did not require recA gene product. Does this mean that there is something very special about plasmid recombination as opposed to bacterial recombination?

S.N. COHEN: These transposable elements, both the IS elements and Tn antibiotic resistance elements, are capable of translocation into the chromosome as well as into plasmids in the absence of recA gene function. They also appear to be implicated in Hfr formation involving interaction between plasmids and chromosomes, so there appears to be nothing unique about plasmids per se so far as these elements are concerned. However, plasmids provide a convenient system for studying illegitimate recombination. I should also mention that Peter Kretschmer in my laboratory has investigated frequencies of translocation of the Tn3 element from plasmid to plasmid, from plasmid to chromosome, and from chromosome to plasmid - and there clearly is a much higher frequency of plasmid-plasmid translocation than translocation involving the chromosome. The reasons for this are not clear; it may be simply that the extrachromosomal state of plasmids provides more opportunity for genome-genome interaction, but we don't know that.

THE EXPRESSION OF YEAST DNA IN ESCHERICHIA COLI

J. CARBON, B. RATZKIN, L. CLARKE, and D. RICHARDSON
Department of Biological Sciences
University of California
Santa Barbara, California 93106 U.S.A.

Abstract: Specific segments of yeast (*S. cerevisiae*) DNA, when cloned on a ColE1 plasmid vector, are capable of complementing auxotrophic deletion mutations in *E. coli*. Hybrid ColE1-yeast DNA plasmids that complement mutations in *leuB* (β-isopropylmalate dehydrogenase), *hisB* (imidazole glycerol phosphate dehydratase), *trpAB* (tryptophan synthase), and *argH* (argininosuccinate lyase) have been isolated and characterized.

INTRODUCTION

Recent studies have shown that segments of yeast (*S. cerevisiae*) DNA can be functionally expressed when cloned in *E. coli* on phage (1) or plasmid (2,3) vectors. These studies use a sensitive assay for the meaningful expression of cloned eukaryotic DNA, the ability to suppress or complement bacterial auxotrophic mutations *in vivo*. Thus, it is possible to detect even relatively low levels of expression by the relief of a metabolite requirement in the mutant bacterial strain. In addition, a wide variety of *E. coli* strains containing mutations in known genes involved in amino acid and nucleotide biosynthesis are available, so that the generality of any complementation phenomena can be established.

Using the poly(dA·dT) "connector" method (4,5) to join randomly sheared yeast DNA segments to plasmid ColE1 DNA (L_{RI}), we have isolated and characterized hybrid plasmids capable of complementing *argH*, *hisB*, *leuB*, and *trpAB* mutations in *E. coli*. The frequency with which this inter-species complementation is observed suggests that it is not a rare phenomenon.

CONSTRUCTION AND USE OF THE HYBRID PLASMID BANK

In order to set up a definitive test for the ability of any given eucaryotic gene system to be expressed and to complement an auxotrophic mutation in *E. coli*, the efficiency of the cloning procedure used must be high enough to insure that sufficient transformant clones containing hybrid DNA plasmids are obtained to be representative of the entire genome of the organism under study. In addition, it is preferable to use DNA segments produced by random scission (hydrodynamic shear) rather than by restriction endonuclease action, to insure that the desired gene system remains intact on at least a portion of the cleaved DNA segments. For example, previous studies have shown that the use of the poly(dA·dT) "connector" method (4,5) to join randomly sheared E. coli DNA with linear ColE1 DNA yields a preparation that will transform E. coli cells to colicin E1 resistance with high efficiency, thereby establishing a collection of transformants containing hybrid plasmids representative of the entire E. coli genome using only 10-15 μg of annealed DNA (6,7).

A preparation of total yeast DNA (from strain X2180-1Aa, SUC2 mal gal2 CUP1) was sheared to an average size of 8-10 X 10^6 daltons, and 3'-poly(dA) homopolymeric extensions added as previously described (2,3,7). After annealing to poly(dT)-tailed ColE1 (L_{RI}) DNA, the preparation (25 μg DNA) was used to transform various E. coli hsm^+ hsr^- strains to colicin E1 resistance. In one experiment, a total of 190,000 colonies containing about 90,000 unique transformant colonies were obtained by this method. Assuming that the yeast genome is 10^{10} daltons and that the sheared pieces of yeast DNA are 10^7 daltons, then only 4600 transformants would be necessary for a 99% probability that the pool of recombinant plasmids would contain any particular yeast DNA segment (7). Thus, the large pool of transformants insures that we have cloned essentially all of the yeast genome.

This collection was screened for hybrid plasmids capable of complementing E. coli auxotrophic mutations in three ways: (a) by F-mediated transfer of hybrid plasmids to suitable auxotrophic F^- recipients (7) (but note comments below concerning recently revised NIH guidelines); (b) by direct plating on selective plates checking for complementation of markers in the bank strain; and (c) by isolation of mixed plasmid DNA from the entire collection, followed by transformation of suitable auxotrophic strains selecting for the desired complementation.

Recently (September, 1976) the NIH Guidelines for Recombinant DNA Research were modified to specifically prohibit the cloning of foreign DNA in hosts containing wild-type conjugative plasmids (e.g., F). We have therefore discontinued all screening by method (a) above.

COMPLEMENTATION OF leuB MUTATIONS BY YEAST DNA

The C600 strain used as host for the construction of the

hybrid ColE1-yeast DNA plasmid bank contains the leu6 mutation, which has been located in the leuB gene (specifying β-isopropylmalate dehydrogenase) (J. Calvo, personal communication). Extracts prepared from this strain contain no detectable activity when assayed for the oxidative decarboxylation of β-isopropylmalate to form α-keto-isocaproic acid (Richardson & Carbon, unpublished data).

Portions of a suspension of 40,000 JA199 transformant colonies containing hybrid ColE1-yeast DNA plasmids were plated on minimal medium with tryptophan and without leucine and incubated at 30°C. After 3-4 days, large and small Leu^+ colonies appeared at a frequency of 10^{-6}. Several of the small colonies were selected and tested for resistance to colicins E1 and E2 (8). The plates were then further incubated at room temperature for 2 weeks, after which time very small (1-2 mm) colonies appeared. Two of the intermediate size Leu^+ colonies and 15 of the very small ones that were colicin E1-resistant and E2-sensitive were also tested for the ability to donate their Leu^+ character to an F^- leuB strA strain (KL380) on minimal medium containing arginine, methionine, and streptomycin. All except one transferred the Leu^+ character at a high frequency (10^{-3} - 10^{-4} per donor cell). The Leu^+ recipients were also colicin E1-resistant and E2-sensitive, an indication that hybrid ColE1 plasmids had been transferred (8).

Covalently closed, supercoiled plasmid DNA was isolated from the 16 Leu^+ colicin E1-resistant transformants from above. The purified plasmid DNAs were treated with excess endonuclease Eco RI and the digests were fractionated by electrophoresis on 1.2% agarose gels. Four different patterns of Eco RI restriction fragments were observed among this group (Fig. 1). Plasmid DNA (pYeleu10) from the fastest growing Leu^+ transformant gave a unique fragment pattern, quite different from that displayed by any of the slower growing transformants. The majority of the latter group gave the restriction pattern shown by pYeleu12 DNA, although two other plasmid types were obtained, as typified by pYeleu11 and pYeleu17 (Fig. 1). These four plasmids are of different sizes, with molecular weights ranging from 8.9×10^6 to 13.5×10^6 daltons (see Table 1). None of the Eco RI fragments obtained from this plasmid group appear to be identical, although in each case the largest fragment is presumed to contain the 4.2×10^6 dalton ColE1 segment.

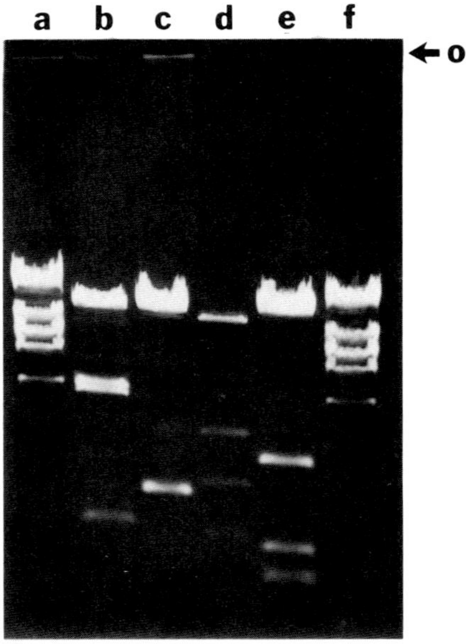

Fig. 1. Fractionation by 1.2% agarose gel electrophoresis of complete endonuclease Eco RI digests of various hybrid ColE1-yeast DNA plasmids; lanes a and f, phage λ DNA; lane b, pYeleu10 DNA; lane c, pYeleu11 DNA; lane d, pYeleu12 DNA; lane e, pYeleu17 DNA. Each well contained 1 μg of DNA predigested with excess endonuclease Eco RI as previously described (5). Electrophoresis was carried out at 8 volts/cm for 2 hours and the DNA visualized with ethidium bromide. The sizes of the λ fragments are (from the top, in daltons): 13.7×10^6, 4.7×10^6, $3.7\text{-}3.5 \times 10^6$, 3.0×10^6, and 2.1×10^6.

These plasmid DNAs (pYeleu10, pYeleu11, pYeleu12, and pYeleu17) were used to transform three different *E. coli leuB* mutants, selecting for either Leu$^+$ or colicin E1 resistance. As shown in Table 1, plasmid pYeleu10 DNA is capable of transforming all of the *leuB* mutants (*leuB6*, *leuB61* and *leuB401*) (11) to both Leu$^+$ and colicin E1 resistance with high frequency (> 10^4 transformants/μg). Although the other three plasmid DNAs would readily transform all of the *leuB* strains to colicin E1 resistance, these plasmids would complement only the mutation in *leuB6* (Table 1). pYeleu10 DNA is not capable of transforming to Leu$^+$ strains bearing mutations in other genes of the *leu* operon, such as *leuA371*, *leuC222*, or *leuD211*

(9).

TABLE 1

Transformation of E. coli auxotrophs by ColE1-yeast hybrid plasmid DNAs

Plasmid DNA	Molecular weight	Strain transformed	Transformants/μg DNA	
			Colicin E1-resistant	Selective marker
pYeleu10	13.5×10^6	JA194(leuB6)	2.7×10^5	3.1×10^5(Leu$^+$)
		CV514(leuB401)	2.1×10^4	2.4×10^4
		CV516(leuB61)	2.6×10^4	1.6×10^4
pYeleu11	8.9×10^6	JA194(leuB6)	7×10^4	3.9×10^3
		CV514(leuB401)	1.4×10^5	0
		CV516(leuB61)	4.2×10^4	0
pYeleu12	10.3×10^6	JA194(leuB6)	7×10^4	7.6×10^3
		CV514(leuB401)	1.7×10^4	0
		CV516(leuB61)	8.8×10^3	0
pYeleu17	13.2×10^6	JA194(leuB6)	7×10^4	7.5×10^3
		CV514(leuB401)	7.3×10^4	0
		CV516(leuB61)	1.8×10^4	0
pYehis1	10.7×10^6	hisB463	1.5×10^4	1.3×10^3(His$^+$)
pYehis2	10.7×10^6	hisB463	2.6×10^4	1.7×10^3
pYehis3	10.7×10^6	hisB463	2.6×10^4	1.9×10^3
pYetrp1	14.9×10^6	trpA36	4.5×10^5	8×10^4(Trp$^+$)
		trpA38	ND	6.6×10^3
		trpB9579	ND	4.6×10^3
		trpAB17	ND	1.8×10^4
pYearg1	14×10^6	JA209(argH)	ND	6×10^3(Arg$^+$)

These data have been corrected for the presence of both spontaneous Leu$^+$ revertants and colicin E1-resistant mutants. Molecular weights were determined by electron microscopy using ColE1 DNA (4.2×10^6 daltons) as standard. ND = not determined.

If pYeleu10 carries a segment of yeast DNA that specifies the synthesis of a functional β-isopropylmalate dehydrogenase, it should complement deletions of the leuB region in the bacterial host cell. Although well-characterized leuB deletions in E. coli K12 were not available, several leu deletion mutants in Salmonella typhimurium have been mapped by Calvo and Worden (10). The pYeleu10 plasmid was transferred from strain JA199 into the *Salmonella* deletion mutants, *leuA124*, *leuB698*, *leuC5076*, and *leuD657* (obtained from Dr. Joseph Calvo), by F-mediated transfer. Leu^+ *Salmonella* colonies were obtained only from the JA199/pYe*leu*10 X *leuB698* cross, with a frequency of 10^{-3} Leu^+ recipients per donor cell. Plasmid DNA isolated from four of the *Salmonella* Leu^+ strains gave an *Eco* RI restriction fragment pattern identical to that from authentic pYe*leu*10 DNA. Thus, the presence of pYe*leu*10 DNA correlates with the transfer of a Leu^+ phenotype to a strain harboring a deletion in *leuB*.

In order to prove that pYe*leu*10 contains a segment of yeast DNA, we labeled the plasmid DNA by nick translation with DNA polymerase I and ^{3}H-labeled deoxynucleoside triphosphates (11), and determined if single-stranded plasmid DNA would reassociate with authentic yeast DNA. The reassociation of single-stranded pYe*leu*10 DNA was driven well by yeast DNA single strands, but not by single-stranded *E. coli* DNA (2,3).

If one assumes that 70% of pYe*leu*10 is yeast DNA (from the total pYe*leu*10 molecular weight, 13.5×10^6, minus the ColE1 segment, 4.2×10^6) then we can calculate from the initial slopes of the reassociation rate curves (2,3) that the yeast segment in pYe*leu*10 is 0.11 - 0.12% of the yeast genome. If we further assume one copy of the pYe*leu*10 segment in the yeast genome, then the total yeast genome would be 7.8 - 8.5 $\times 10^9$ daltons, well within the range of values in the literature ($5 - 10 \times 10^9$ daltons) (12). Apparently this segment of DNA is a unique fragment of the yeast genome.

COMPLEMENTATION OF *hisB* MUTATIONS

Mixed plasmid DNA isolated from the 40,000 transformant colonies was used in attempts to transform several *E. coli* auxotrophic strains to prototrophy. Among these strains was *hisB463*, which lacks an active imidazole glycerol phosphate dehydratase. Struhl *et al.* (1) have previously isolated a lambda phage containing a segment of yeast DNA (λgt-Sc*his*) that suppresses this mutation. Two micrograms of our mixed hybrid plasmid DNA was used to transform strain *hisB463*, selecting for His^+. Three His^+ colonies were detected after 5 days of incubation at 30°, as compared to 10^5 colicin E1-resistant colonies. These colonies were colicin E1-resistant, but sensitive to colicin E2. Plasmid DNAs were isolated from these strains and used to transform strain *hisB463* again. As shown in Table 1, these purified plasmid DNAs (pYe*his*1, pYe*his*2 and pYe*his*3) transformed the *hisB463* strain to colicin E1-resistance and to His^+ with high frequency. Transformation to His^+ occurred at a lower frequency than to colicin E1-

resistance, although prolonged incubation produced more His$^+$ colonies.

All three pYe*his* plasmids are of similar size, with a molecular weight of 10.7 X 10^6 daltons. They each contain a single *Eco* RI endonuclease restriction site within the segment of cloned yeast DNA, and are probably identical.

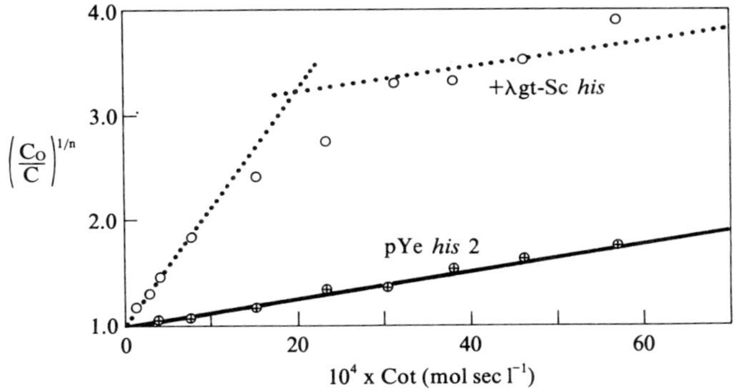

Fig. 2. Reassociation kinetics of labeled single stranded pYe*his*2 DNA fragments (0.07 μg/ml) in the presence of single stranded fragments of: (⊕——⊕) salmon sperm DNA, 1000 μg/ml; and (O---O) λgt-Sc*his* DNA, 125 μg/ml. Plasmid DNA was labeled by nick translation, and rehybridization kinetics were determined as previously described (11).

Figure 2 shows the reassociation kinetics of single stranded pYe*his*2 DNA in the presence or absence of single stranded λgt-Sc*his* DNA, from the recombinant λ phage containing a segment of yeast DNA that suppresses the *hisB463* mutation (1). If the pYe*his*2 plasmid and λgt-Sc*his* contain similar segments of yeast DNA, then the addition of λgt-Sc*his* single stranded DNA should increase the rate of reassociation of single stranded pYe*his*2 DNA. As shown in Fig. 2, λgt-Sc*his* DNA markedly increases the rate of reassociation of pYe*his*2 DNA. The fraction of λgt-Sc*his* DNA that is homologous to the pYe*his*2 DNA can be calculated as above to be equal to 22% of the λgt-Sc*his* DNA (30 X 10^6 daltons), equivalent to 6.6 X 10^6 daltons of DNA. From the molecular weights of pYe*his*2 and Col E1 DNAs and assuming a full-length ColE1 segment is present in pYe*his*2, the calculated size of the cloned yeast DNA segment would be 6.7 X 10^6 daltons. pYe*his*2 DNA contains a single *Eco* RI site, which is apparently identical to one of the two sites defining the yeast segment in λgt-Sc*his* DNA.

COMPLEMENTATION OF *trpAB* (TRYPTOPHAN SYNTHASE) MUTATIONS

Collections of colonies bearing hybrid ColE1 plasmids can also be screened by F-mediated transfer to suitable auxotrophic recipients (7). A set of 4300 F^+ hybrid ColE1-yeast DNA colonies were mated with strain CH754 (F^- *argH metE recA56 trpA36 xyl*) by the plate mating technique previously described (7), selecting for transfer of the Trp^+ phenotype. One colony from the collection was capable of transferring the Trp^+ character to F^- *trpA* or *trpB* recipients.

Purified plasmid DNA (pYe*trp1*) from this strain has a molecular weight of 14.9×10^6 daltons. As shown in Table 1, this DNA can transform both *trpA* and *trpB* mutants to Trp^+ and colicin E1 resistance with high frequency. pYe*trp1* DNA complements not only missense (*trpA36*) and nonsense (*ochre*) (*trpA38* and *trpB9579*) mutations, but also complements a deletion of the entire *trpAB-tonB* region (*trpAB17*). It seems likely, therefore, that pYe*trp1* specifies the synthesis of a completely functional tryptophan synthase (*trpA* and *trpB* specify the non-identical α and β subunit proteins of the *E. coli* tryptophan synthase).

Plasmid pYe*trp1* was shown to contain a segment of cloned yeast DNA by measuring the rates of reassociation of labeled single-stranded plasmid DNA in the presence of excess unlabeled single-stranded DNAs from salmon sperm, *E. coli*, or yeast, as described previously for pYe*leu10* and pYe*his* DNAs (2,3). In the presence of unlabeled single-stranded yeast DNA (80 μg/ml), the initial rate of reassociation of single-stranded pYe*trp1* DNA (0.01 μg/ml) was increased about 5.3-fold; however, unlabeled single-stranded *E. coli* DNA (80 μg/ml) did not affect the pYe*trp1* reassociation. Thus, we conclude that the cloned segment of DNA in pYe*trp1* is indeed derived from yeast.

COMPLEMENTATION OF *argH* MUTATIONS

The *argH* locus in *E. coli* specifies the synthesis of argininosuccinate lyase, the last enzyme in the pathway of arginine biosynthesis (13). A preparation of mixed hybrid ColE1-yeast DNA (10 μg) isolated from 4300 transformant colonies was used to transform *E. coli* strain JA209 (*argH metE recA str trpA*), selecting for Arg^+ and (on separate plates) colicin E1 resistance. This strain has never been seen to revert spontaneously to Arg^+ in our hands. The mixed hybrid plasmid DNA transformed strain JA209 to colicin E1-resistance with an efficiency of 4×10^4 transformants/μg. One Arg^+ colicin E1-resistant colony was obtained from this experiment.

After colony purification, plasmid DNA (pYeargl) was isolated from this clone and purified by CsCl-ethidium bromide banding. As shown in Table 1, the purified pYeargl DNA (MW = 14 X 10^6) transformed strain JA209 (argH$^-$) to Arg$^+$ with high frequency (6 X 10^3 transformants/μg DNA). The exact nature of this complementation is currently under investigation.

DISCUSSION

We have isolated several hybrid ColE1-yeast DNA plasmids that are capable of complementing various auxotrophic mutations in E. coli. The plasmids were constructed by joining sheared yeast DNA to plasmid ColE1 DNA using poly(dA·dT) "connectors". Most of these plasmids are in the 10-15 X 10^6 daltons size range.

The relative ease with which we have identified hybrid plasmids capable of complementing E. coli mutations suggests that meaningful expression of yeast DNA in bacteria is not a rare phenomenon. For example, out of 15 complementations attempted with our hybrid plasmid collection, four (argH, hisB, leuB, and trpAB) were successful. Thus far, we have been unsuccessful in attempts to complement mutations in the following E. coli genes: galKT, hisC, hisD, hisF, ilvA, ilvC, ilvE, metE, pyrB, thyA, trpE (possible complementation of hisF and metE has been observed, but not verified as yet).

The actual mechanism of complementation by the cloned yeast DNA is not clear. The hisB463 mutation that is complemented by the yeast DNA segment in pYehis and λgt-Schis is thought to be a deletion (1). The pYeleu10 plasmid is capable of complementing leuB deletion mutations in S. typhimurium. Similarly, the pYetrp1 plasmid complements a total deletion of the trpAB region in E. coli. Thus, in these cases, we are relatively certain that a functional enzyme is being synthesized from genetic information on the hybrid plasmid. We do not yet know, however, if the new enzymes synthesized in response to these hybrid plasmids are identical to the corresponding enzymes from yeast.

Extracts prepared from the host leuB6$^-$ E. coli strain, from the same strain bearing the pYeleu10 plasmid, and from leuB$^+$ E. coli and leu2$^+$ yeast strains have now been assayed for β-isopropyl malate dehydrogenase activity (Richardson, Ratzkin, and Carbon, unpublished data). The specific activity of the enzyme in the E. coli leuB6$^-$/pYeleu10 extracts is about one-fifth that of extracts prepared from the corresponding leuB$^+$ strain. This activity is undetectable in extracts from the parent leuB6$^-$ strain lacking the plasmid. It is noteworthy that the activities in both leuB6⁻/pYeleu10 and in yeast extracts are extremely cold-sensitive, being inactivated after only 20 minutes at 0°. However, the enzyme in wild-type *E. coli* leuB+ extracts is still quite active after prolonged storage at 0°. Although these results are still preliminary, it appears that the dehydrogenase specified by pYeleu10 is quite similar in its properties to the β-isopropyl malate dehydrogenase present in yeast extracts. A final judgment on the exact nature of the enzyme specified in *E. coli* by pYeleu10 DNA must await purification and characterization of both this enzyme and the one from yeast cells.

Certainly the mechanism of this type of complementation or suppression can differ in various cases. For example, in addition to pYeleu10, three other hybrid ColE1-yeast DNA plasmids (pYeleu11, pYeleu12, pYeleu17) have been found which suppress the leuB6 mutation, although with a much lower efficiency than does pYeleu10. These plasmids all display different patterns of cutting by endonuclease *Eco* RI. Although pYeleu10 will strongly complement any of the leuB mutations we have tried (Table 1), these other pYeleu plasmids appear to suppress only the leuB6 mutation in C600 strains. The possibility exists of missense suppression mediated by yeast tRNA or aminoacyl-tRNA synthetases produced from yeast genes on the plasmid, since it is well known that tRNA mis-aminoacylation occurs readily in *E. coli*-yeast heterologous systems (14). Recently, however, we have determined that the leuB6-specific plasmid DNAs do not hybridize to ^{32}P-labeled yeast tRNA, an indication that tRNA genes are absent on these plasmids (A. C. Chinault, unpublished data).

The nature of the transcription and translation signals utilized during RNA and protein synthesis specified in *E. coli* by the hybrid ColE1-yeast DNA plasmids is still unclear. This aspect of the problem is currently under active investigation in our laboratories.

ACKNOWLEDGMENTS

We are grateful to Gary Tschumper and Margaret Nesbitt for valuable technical assistance; to K. Struhl and R. Davis for a gift of λgt-Schis DNA; and to J. Calvo and C. Yanofsky for supplying several key leu and trp mutant strains. This work was supported by research grants CA-11034 and CA-15941 from the National Cancer Institute, and by a research grant from Abbott Laboratories.

REFERENCES

(1) K. Struhl, J. R. Cameron and R. W. Davis, Proc. Nat. Acad. Sci. U.S.A. 73 (1976) 1471-1475.

(2) B. Ratzkin and J. Carbon, Proc. Nat. Acad. Sci. U.S.A., in press.

(3) J. Carbon, L. Clarke, C. Ilgen and B. Ratzkin, Miles Symposia Proceedings, Tenth International Symposium (Raven Press, New York, 1976) in press.

(4) P. E. Lobban and A. D. Kaiser, J. Mol. Biol. 78 (1973) 453-471.

(5) D. A. Jackson, R. H. Symons and P. Berg, Proc. Nat. Acad. Sci. U.S.A. 69 (1972) 2904-2909.

(6) L. Clarke and J. Carbon, Proc. Nat. Acad. Sci. U.S.A. 72 (1975) 4361-4365.

(7) L. Clarke and J. Carbon, Cell 9 (1976) 91-99.

(8) V. Hershfield, H. W. Boyer, C. Yanofsky, M. A. Lovett and D. R. Helinski, Proc. Nat. Acad. Sci. U.S.A. 71 (1974) 3455-3459.

(9) J. M. Somers, A. Amzallag and R. B. Middleton, J. Bacteriol. 113 (1973) 1268-1272.

(10) J. M. Calvo and H. E. Worden, Genetics 64 (1970) 199-214.

(11) F. H. Schachat and D. S. Hogness, Cold Spring Harbor Symp. Quant. Biol. 38 (1973) 371-381.

(12) G. D. Lauer and L. C. Klotz, J. Mol. Biol. 95 (1975) 309-326.

(13) W. K. Maas, R. Maas, J. M. Wiame and N. Glansdorff, J. Mol. Biol. 8 (1964) 359.

(14) K. B. Jacobson, in: Progress in Nucleic Acid Research and Molecular Biology, Vol. 11, eds. J. N. Davidson and W. E. Cohn (Academic Press, New York, 1971) pp. 461-488.

DISCUSSION

P. BERG: I would like to comment on a point Charles Weissmann raised earlier, concerning the problem of evolutionary drift in the cloned segments. It is quite clear that when you can select or score for a marker you can pick up charges in its structure quite readily as changes in the growth rate if the cells are dependent upon the level of the inserted segment's activity. On the other hand, many people will be cloning segments for which they have no selection and no way of detecting a specific activity or function. I think it is important to bear in mind that the segment that is eventually cloned may well be a skeleton of what the original sequence in the chromosome is, and that it is important to devise methods that prove that a particular segment from genome X has not undergone drastic changes e.g., deletions, rearrangements, insertions. It will be necessary to prove that the cloned segment is identical to the corresponding segment in the chromosome.

J. CARBON: Yes, you would need to go back and use Southern gels to clone the same segment of DNA by a method that does not involve selection for complementation, and then determine the homology between what we have cloned and what is actually in the DNA.

P. BERG: Yes, but point mutations are not likely to be detected by changes in gel patterns. If you drop your selection procedure you may pick up mutations. That is the problem I am pointing out.

J. CARBON: Another point I might make with regard to that, Paul; we can grow these cultures under non-selective conditions for many generations and then isolate plasmid. The DNA will still transform with very high frequency to the desired genotype. So you don't really need to maintain selective pressure to keep the main population of plasmid intact and functional.

P. BERG: I think we must distinguish between the results obtained from examining the entire population and looking at individual molecules. A subclone from a population of recombinant plasmids for a hundred generations have mutations. Such mutations may not be detectable in the total population because they are averaged out over many different individuals.

F. BLASI: I would like to comment on your histidine yeast plasmid. The *E. coli* strain that you are using for complementation has been used by us for quite a few years, though we have never assayed the dehydratase activity. I know it is certainly *hisC*⁻, a transaminase mutant.

J. CARBON: Well, actually Ron Davis may be able to answer that better than I. They have made extracts from the *hisB463* strain and looked at enzyme levels. They have published that it lacks the dehydratase activity. We haven't done that in our laboratory so I can not answer this.

F. BLASI: Well, I never assayed the dehydratase but I certainly know it lacks the transaminase.

J. CARBON: I think they were told originally that the strain was *hisC*⁻, but when they did the enzyme assays they discovered it was lacking *HisB* activity rather than *hisC*.

R.B. MEAGHER: What *Eco* R1 fragments were required for complementationof the tryptophan auxotrophy. What were the molecular weights, and what was the total amount of DNA required for complementation?

J. CARBON: Three *Eco*R1 fragments totaling about 3.8 million daltons.

R.B. MEAGHER: Have you been able to get any smaller fragments than that to complement?

J. CARBON: We are trying to get smaller pieces of DNA, but at the moment the smallest segment that will still complement is composed of those 3 R1 fragments that sum up to about 3.8 million daltons of DNA. The smallest one couldn't be more than half a million.

R.B. MEAGHER: The enzyme wouldn't require anywhere near that much, would it? I don't know how large the protein subunits are.

J. CARBON: Well, there is one paper on purification of yeast tryptophan synthetase. The subunit was reported to be about 70,000 daltons.

S.B. GREER: Bruce Carlton showed by using assembly mutants in *B. subtilis* that the tryptophan synthesizing enzymes appear to be part of a multienzyme complex. Is it known whether this is also true in *E. coli*? If so, the complementation would be even more remarkable.

J. CARBON: The *E. coli* enzyme is known to contain four subunits, two α and two β, but it is not known, as far as I know, that it exists in a multienzyme complex with several other enzymes. In the case of the yeast enzyme, it is also not known.

CHARACTERIZATION OF TETRACYCLINE AND AMPICILLIN RESISTANT PLASMID CLONING VEHICLES

R.L. RODRIGUEZ, R. TAIT, J. SHINE, F. BOLIVAR,
H. HEYNEKER, M. BETLACH AND H.W. BOYER
Department of Biochemistry and Biophysics
University of California, San Francisco
San Francisco, California 94143

Abstract: Our laboratory has been constructing plasmids with suitable genetic properties for the cloning of DNA fragments. We have combined the essential replication properties of the plasmid ColE1 with two antibiotic resistance markers (tetracycline and ampicillin) to generate a series of related plasmids. During the course of the construction of these plasmids we have determined the positions of restriction sites used for cloning DNA fragments relative to the antibiotic resistance genes and the origin of DNA replication. We have mapped a number of restriction sites on a series of related plasmids. We will summarize the development of two plasmids, pBR313 and pBR322, which are useful cloning vehicles for *in vitro* recombinant DNA research. The characterization of these plasmids has prompted an investigation of the nature of tetracycline resistance determined by the plasmid pSC101 and the pMB9 series of plasmids.

INTRODUCTION

There are several essential steps in *in vitro* recombinant DNA technology required for the insertion of DNA fragments from any source into either viral or plasmid replicons. Generally speaking, these steps involve the specific cleavage of the desired DNA with restriction endonucleases (1), the covalent rejoining of DNA fragments to the plasmid cloning vehicle (2), and the transformation of a suitable bacterial host strain, e.g., *Escherichia coli*, K12, with recombinant DNA molecules (3). The recovery of bacterial clones containing replicating recombinant plasmids can be readily obtained with this procedure. We feel that the most critical component of this technology is the cloning vehicle.

The degree of usefulness of any plasmid cloning vehicle can be determined by such criteria as the ease of plasmid isolation, the presence of readily detectable genetic markers and the number and location of unique restriction endonuclease cleavage sites. However, it is the ease with which recombinant plasmids can be differentiated from non-recombinant plasmids, i.e. plasmid vehicles, that is of primary importance to most recombinant DNA researchers. We have previously reported (4, 5) the use of recombinant DNA technology to construct a number of ampicillin (Ap) and tetracycline (Tc) resistant plasmids which meet all of the above mentioned criteria for efficient cloning vehicles. In addition, these plasmids exhibit a high level of containment with the *E. coli* host strain X1776 (6) according to the requirements for an EK2 host vector system. This paper briefly describes some salient features of two cloning vehicles, pBR313 and pBR322, as well as summarizes our current understanding of the genetic components involved in the expression of Tc resistance. The Tc resistance mechanism specified by pBR313 and pBR322 reportedly originates from the plasmid R6-5 (7) via the plasmids pSC101 and pMB9 respectively (4). We believe that Tc resistance in pMB9 and the pBR series plasmids is only part of the normal, inducible (8) Tc resistance mechanism expressed by pSC101.

EXPERIMENTAL PROCEDURES

(a) Bacterial Strains:

The following derivatives of *E. coli* K12 were used as host strains for the plasmids pBR313 and pBR322: HB101 F^- *pro leu thi lacy* *str*^{r} $r_K^- m_K^-$ endoI^-, *recA*$^-$; RR1 F^- *pro leu thi lacy* *str*^{r} $r_K^- m_K^-$; endoI^- (5). Plasmids were also examined in the "minicell" strain of *E. coli*, P5678-54 thr^- leu^- B_1^- supE lacY tonA gal^- mal^- xyl^- ara^- mtl^- min^- (9).

(b) Materials

Salts, media, sugars, and buffers were purchased from J. T. Baker Chemical Company, Difco Corp., Mallinckrodt Corp., and Sigma Chemical Company. Tetracycline was purchased from Sigma Chemical Company.

(c) Preparation of plasmid DNA:

Preparation of plasmid DNA by amplification in the presence of chloramphenicol (170 μg/ml) was performed according to Clewell (10). The DNA was purified by a modification of the cleared lysate technique of Guerry *et al.* (11). Plasmid copy number was determined by the sarkosyl lysate method as previously described (12).

(d) Minicell preparations:

An analysis of plasmid specified Tc^r proteins in the minicell system was performed according to the procedure described by Meagher et al. (13).

(e) Enzymes and reaction conditions:

All restriction enzymes described in the text were purified according to a procedure described by Greene et al. (14). T4 polynucleotide ligase was purified according to the procedure of Panet et al. (15). E. coli DNA-dependent, RNA polymerase and S1 nuclease were gifts from Michael Chamberlain and A. Dugaiczyk respectively. Reaction conditions for the restriction enzymes, T4 polynucleotide ligase and S1 nuclease have been described elsewhere (5). The RNA polymerase binding reaction was performed according to Seeburg and Schaller (16).

(f) Agarose electrophoresis:

The procedure for agarose slab gel electrophoresis has been reported previously (5).

RESULTS AND DISCUSSION

pBR313 and pBR322 plasmid cloning vehicles:

The plasmid pBR313 combines the replication elements and colicin E1 immunity gene of a plasmid very similar to ColE1 with the Tc resistance genes of pSC101 and the Ap resistance gene of pSF2124 (4, 5). Neither antibiotic resistance marker can be transposed (5) to other plasmids. This plasmid has a molecular weight of 5.8 megadaltons, and more than forty restriction sites cleaved by thirteen different restriction endonucleases have been located on the molecule (Fig. 1a). At least fourteen of these cleavage sites have been mapped in the Tc resistance locus. There are single cleavage sites for the EcoRI, HindIII, BamI, SalI, SmaI and HpaI endonucleases on pBR313 which can be used for cloning fragments of DNA generated by these enzymes. These six restriction endonucleases can be used in fourteen different combinations to subclone smaller components of larger DNA fragments previously cloned in one of the unique sites of pBR313. This feature of pBR313 is of particular importance for the sequencing technique recently developed by Maxam and Gilbert (17) which calls for substantial amounts of highly purified low-molecular

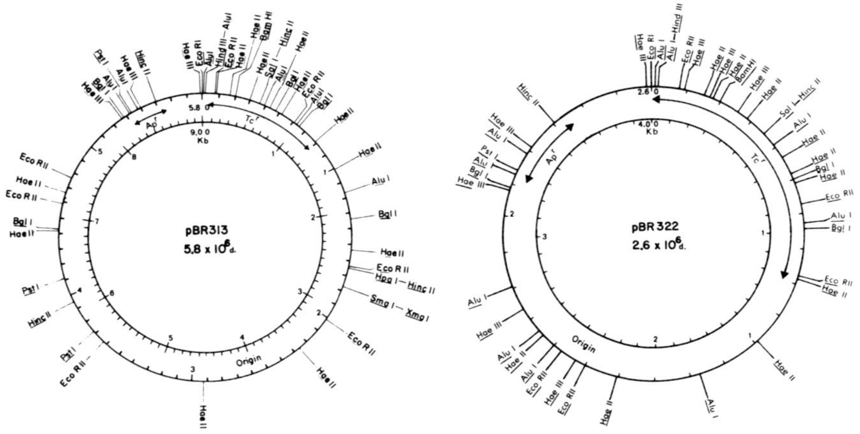

Fig. 1. The circular restriction maps of (a) pBR313 and (b) pBR322. The relative positions of restriction sites are drawn to scale on a circular map divided into units of 1×10^5 daltons (d) and 0.1 kilobases (Kb).

weight DNA fragments. Molecular cloning experiments have revealed that the HindIII, SalI and BamI cleavage sites are located in the Tc resistance locus. This feature facilitates the screening of transformants for recombinant plasmids by scoring Ap resistant transformants for a Tc sensitive phenotype. Cultures in which greater than 90% of the cells harbor recombinant plasmids can be obtained by exploiting the bacteriostatic effect of Tc on Tc sensitive cells (4).

A derivative of pBR313, pBR322, has been constructed which provides several other advantages as a cloning vehicle. The principle reason for constructing this vehicle was to eliminate all but the one PstI cleavage site located in the Ap resistance gene. Fragments of DNA generated by the PstI endonuclease can be cloned in this site and recombinants identified by their Ap sensitive and Tc resistant phenotypes. Moreover, the PstI endonuclease generates DNA fragments with single strand 3'OH termini which serve directly as a primer for calf thymus terminal transferase (6). The cloning of DNA fragments via the construction of complementary deoxyhomopolymeric termini is greatly simplified with the use of PstI digested pBR322 plasmid. Approximately 36 cleavage sites made by 12 different restriction endonucleases have been located on this molecule (Fig. 1b). There is one cleavage site for each of the BamI, EcoRI, HindIII, PstI and SalI restriction endonucleases and two cleavage sites for the HincII restriction endonuclease. The relative positions of these seven were conserved with respect to pBR313 during the construction of pBR322. The cloning of DNA fragments into the BamHI, HincIII, PstI and SalI restriction sites inactivates one or the other antibiotic resistance marker. The molecular weight of pBR322 has been substantially reduced to 2.6 megadaltons. As a result, pBR322 does not carry a functional colicin E1 immunity gene or the SmaI and HpaI cleavage sites present in pBR313.

An important objective in our construction of plasmid cloning vehicles was to combine and maintain the replication elements of the colE1-like plasmid, pMB1, with the Ap resistance and Tc resistance genes from pSF2124 and pSC101 respectively-

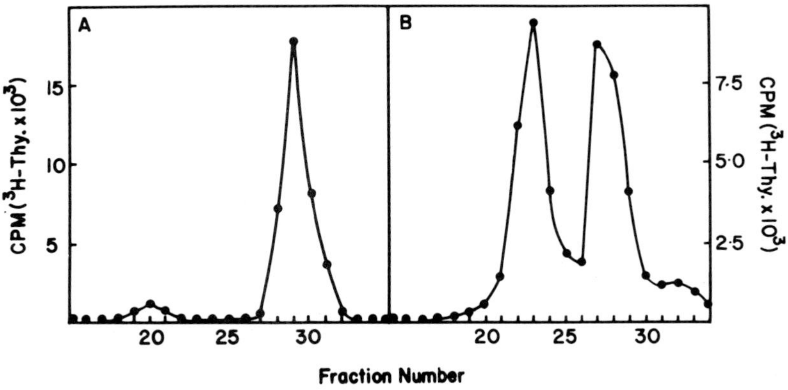

Fig. 2. Sarkosyl lysate dye-buoyant density centrifugation of ^{3}H-thymine labeled pBR313 containing cells before (a) and after (b) 14 hours of incubation in the presence of chloramphenicol. The plasmid peaks at fractions 20 (A) and 23 (B) represent 4.2% and 48% of the total cellular DNA respectively.

tively. Figure 2(a and b) shows that pBR313 exhibits a "relaxed" mode of DNA replication like that previously reported for ColE1 (10). In the presence of chloramphenicol at concentrations which inhibit gross protein synthesis, pBR313 amplifies the number of plasmid molecules from 50 to 420 per cell. An analysis of the number of pBR322 molecules/cell gives similar results (unpublished observation). The amplifiable feature of relaxed-replicating plasmids such as pBR313 and pBR322 has proven to be of tremendous importance to recombinant DNA technology.

Tetracycline resistance:

As pointed out above, the relative positions of the HindIII, BamHI and SalI cleavage sites in the pMB9-pBR322 series of plasmids are the same in pSC101. Cloning of DNA fragments in one of these sites can be detected by screening transformants for a Tc sensitive phenotype. We (and others) have found that not all DNA fragments cloned in the HindIII site of pMB9, pBR313 or pBR322 result in Tc sensitivity. This prompted us to study in more detail the Tc resistance determined by pSC101 and the pMB9 series of plasmids.

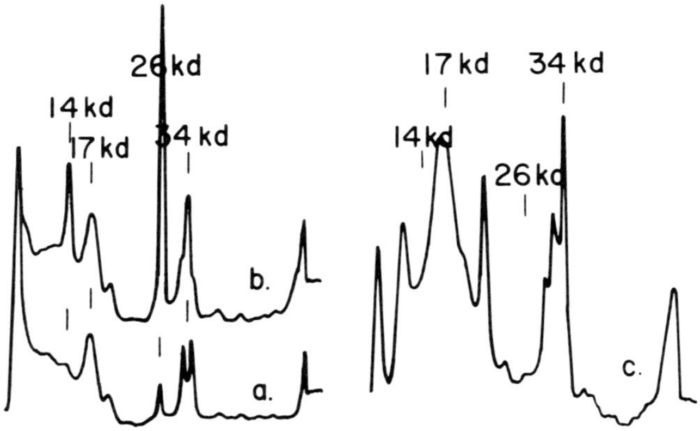

Fig. 3. Polypeptides synthesized in minicells containing pSC101 and pMB9. Densitometric scan of radioactive polypeptides produced by minicells containing: a) pSC101 labeled in the absence of tetracycline, b) pSC101 labeled in the presence of 5 μg/ml tetracycline, c) pMB9 labeled in the absence or presence of tetracycline.

The plasmid pSC101 codes for an inducible Tc resistance mechanism (8). One component of this mechanism is a basal, uninduced level of resistance which can be elevated to accomodate higher levels (two-fold) of the antibiotic by a short exposure to low levels of Tc. The pattern of polypeptides synthesized by minicells containing pSC101 is influenced by exposure to Tc, with a five to eight-fold increase in synthesis of three polypeptide species in the presence of the antibiotic (Fig. 3). Investigation of the uptake of ^{3}H-Tc by sensitive cells reveals a rapid influx and efflux of Tc followed by a slower influx of antibiotic (Fig. 4). The rapid

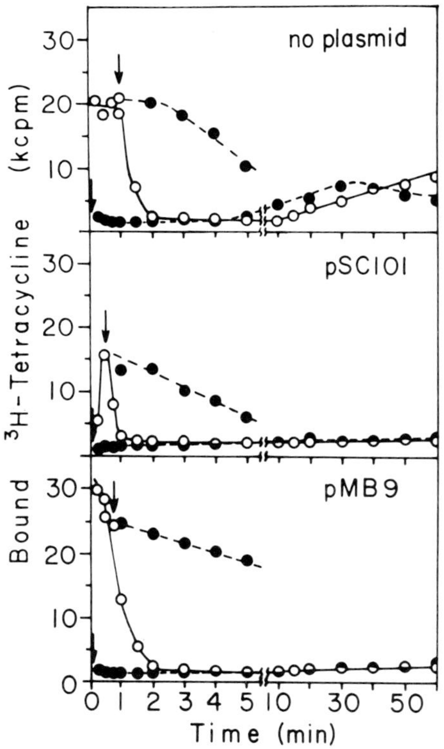

Fig. 4. Binding of ^{3}H-tetracycline. Cultures were grown at 37°C in LB medium to an OD_{650}=0.2±0.02 prior to the addition of ^{3}H-tetracycline at a concentration of 1 μg/ml. Arrow denotes the time of addition of 5 mM sodium arsenite to inhibit ATP production. Binding was measured by pipetting 1 ml aliquots into 2 ml of iced 0.14 M NaCl, collecting the cells by filtration, and washing each filter with 5 ml of cold 0.14M NaCl prior to determination of radioactivity (Connamacher, Mandel and Hahn, 1967). Bound tetracycline —O—; tetracycline bound in the presence of sodium arsenite —●— .

influx and rapid efflux are ATP dependent, while the slow influx is ATP independent. The presence of the plasmid pSC101 results in a substantial decrease in the initial rapid influx and prevents the slow influx of Tc by a mechanism that is independent of ATP. The cloning of DNA fragments at the EcoRI site of pSC101 also significantly effects the level of Tc resistance, the inducible phenotype, (8) as well as the kinetics of Tc uptake, and the pattern of polypeptides observed in minicells (unpublished observations).

It has become apparent, that during the construction of pMB9, the integrity of the Tc resistance mechanism of pSC101 was not maintained. Cells containing pMB9 are constitutively resistant to Tc concentrations as high as 80 μg/ml, nearly three times the maximum resistance level of cells containing pSC101. However, it has been recently reported that when the Tc resistance gene of pBR313 (see Fig. 1a) is integrated into a plasmid which is present in only a few copies per cell (e.g., an F factor) the plasmid confers resistance to only 5μg/ml of Tc (18). Therefore, it is likely that the high level Tc resistant phenotype of pBR313 is merely the expression of multiple copies of a low level Tc resistance gene. Minicells containing the pMB9 plasmid appear to synthesize only two of the pSC101 polypeptides involved in Tc resistance (Fig. 3). Unlike pSC101, cells containing pMB9 exhibit an initial rapid influx and efflux of Tc similar to sensitive cells, although the ATP independent mechanism continues to prevent the slow influx of the antibiotic (Fig. 4). This suggests that the resistance mechanism maintained on the pMB9 series of plasmids functions by preventing the slow accumulation of Tc, while the additional genes present on pSC101 but not pMB9 are involved in the alteration of the initial rapid influx of Tc.

Cells containing recombinant plasmids of pBR313 with DNA fragments cloned in the HindIII site are usually sensitive to a Tc concentration of 0.7 μg/ml in liquid culture. However, the minimum inhibitory concentration of Tc for cells containing certain recombinant plasmids may be measurably higher. This observation leads us to believe that the HindIII site is not situated in a structural gene of the Tc resistance locus. In order to explore this possibility we initiated an analysis of the nucleotide sequence of the HindIII site of pBR313 (Fig. 5). The HindIII cleavage site is located at the terminus of a rather significant region of two-fold symmetry (16 nucleotides with a three nucleotide hyphenation). Two AluI endonuclease cleavage sites which have the internal tetranucleotide sequence recognized by the HinIII endonuclease flank the ends of the symmetrical sequences. The first 100 nucleotides of the sequence contains stop codons in both strands in all phases. At position 106-108 there is an F-met

```
     HaeIII    10    EcoRI 20        30   AluI 40 TaqI HindIII
5' - GGCCCTTTCGTCTTCAAGAATTCTCATGTTTGACAGCTTATCATCGATAAGC -
3' - CCGGGAAAGCAGAAGTTCTTAAGAGTACAAACTGTCGAATAGTAGCTATTCG -

            60         70         80         90        100
   - TTTAATGCGGTAGTTTATCACAGTTAAATTGCTAACGCAGTCAGGCACCGTGT -
   - AAATTACGCCATCAAATAGTGTCAATTTAACGATTGCGTCAGTCCGTGGCACA -

        110        120 HhaI  130        140        150 EcoRII
   - ATGAAATCTAACAATGCGCTCATCGTCATCCTCGGCACCGTCACCCTGGATGC -
   - TACTTTAGATTGTTACGCGAGTAGCAGTAGGAGCCGTGGCAGTGGGACCTACG -

     160       170        HpaII
   - TGTAGGCATAGGCTTGGTTATCCGGTA - 3'
   - ACATCCGTATCCGAACCAATAGGCCAT - 5'
```

Fig. 5. Nucleotide sequence of a 185 base pair HaeIII fragment encompassing the EcoRI and HindIII site of pBR313.

codon followed by 28 amino acid codons. It is possible that this represents the amino terminus of one of the polypeptides involved in Tc resistance coded for by this plasmid. Additional nucleotide sequences are being determined in order to obtain more information about the putative structural gene for Tc resistance.

Binding studies involving *E. coli* DNA-dependent RNA polymerase and plasmid DNA have revealed additional insights into the expression of Tc^r in pBR313 and pSC101. RNA polymerase prebound to pBR313 DNA specifically inhibits cleavage of the HindIII site (Fig. 6 slot 4). Furthermore, when the

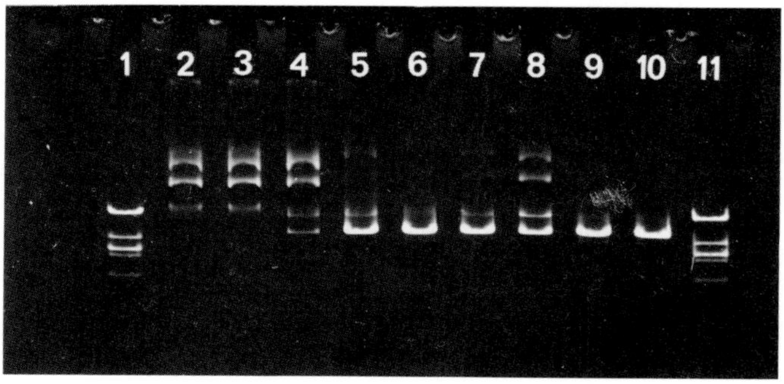

Fig. 6. Agarose slab gel electrophoresis of RNA polymerase (RNAP) binding experiments. Slot 1 and 11; EcoRI lambda DNA markers: slot 2; pBR313 DNA alone: slot 3; pBR313 with RNAP: slot 4; pBR313, RNAP followed by HindIII digestion: slot 5; pBR313, RNAP, 4 ribonucleoside triphosphates followed by HindIII digestion: slot 6; pBR313, HindIII digest alone: slot 7; pBR313 followed by EcoRI digest: slot 8 pBR313, RNAP, followed by EcoRI digest: slots 9 and 10; pBR313, RNAP followed by BamHI and SalI digestion respectively. The RNAP to pBR313 molecular ratio was 2:1.

ribonucleoside triphosphates ATP, CTP, GTP and UTP are added to the reaction mix after RNA polymerase but before the *Hind*III endonuclease, the *Hind*III site is readily cleaved (Fig. 6 slot 5). Additional studies (Fig. 6 slots 9 to 10) show that only the *Hind*III site and to some extent the *Eco*RI site is protected from cleavage by RNA polymerase. The partial protection observed with the *Eco*RI site of pBR313 may be simply due to its proximity to the Tc resistance promoter.

The insertion of nucleotides into the region of the *Hind*III site has been used to study their effect on the level of Tc resistance. The construction of these alterations involve the blunt-end joining (19) of *Hind*III cleaved plasmid DNA. The cohesive *Hind*III end are blunted by either partial (Fig. 7c) or complete (Fig. 7d) S1 nuclease digestion or polymerization with T4 DNA polymerase (Fig. 7b). In some instances (Fig. 7c and d) a synthetic *Eco*RI octanucleotide containing the (20) *Eco*RI site is blunt-end ligated into the *Hind*III site thus converting it to an *Eco*RI site. Blunt-end ligation of the *Hind*III site with or without (Fig. 7b) the *Eco*RI octanucleotide results in the loss of the *Hind*III site. All of these alterations result in a decrease in the level of Tc resistance. This decrease is also accompanied by a decrease in the efficiency of RNA polymerase binding as is evident by the increased polymerase to DNA ratio necessary to protect the *Hind*III (Fig. 7a) and *Eco*RI site (Fig. 7c and d).

	30 40 50 60	Alteration	Tc Resistance	Restriction Site Protection	RNAP: DNA Ratio
A	-TTGACAGCTTATCATCGATAAGCTTTAATGCGGTAGTTT- -AACTGTCGAATAGTAGCTATTCGAAATTACGCCATCAAA-	none	80-100	yes	1:1
B	-TTGACAGCTTATCATCGATAAGCTAGCTTTAATGCGGTA- -AACTGTCGAATAGTAGCTATTCGATCGAAATTACGCCAT- +4	+4 NBP	50	-	-
C	-TTGACAGCTTATCATCGATAAGCTGAATTCAGCTTTAAT- -AACTGTCGAATAGTAGCTATTCGACTTAAGTCGAAATTA- +10	+10NBP	2	yes	20:1
D	-TTGACAGCTTATCATCGATATGAATTCATTAATGCGGTA- -AACTGTCGAATAGTAGCTATACTTAAGTAATTACGCCAT- +4	+4 NBP	10	yes	2:1

Fig. 7. Nucleotide sequences of the Tc resistance promoter region. (A) Unaltered *Hind*III region. *Hind*III site altered by (B) *Hind*III cleavage followed by polymerization T4 DNA polymerase, (C) *Hind*III cleavage followed by partial S1 nuclease digestion and blunt-end ligation of the *Eco*RI octanucleotide, (D) *Hind*III cleavage followed by complete S1 nuclease digestion and blunt-end ligation of the *Eco*RI octanucleotide. Tc resistance is expressed in µg/ml. NBP = nucleotide base pairs.

The evidence presented thus far supports our belief that the HindIII cleavage site lies in or near an RNA polymerase promoter involved in the expression of the Tc resistance mechanism. Furthermore, we feel that the high level resistance to Tc specified by pMB9, pBR313 and pBR322 represents only one component of the native Tc resistance mechanism which was enzymatically dissected from pSC101. This component probably involves the expression of a gene(s) which prevents the slow accumulation of Tc in the cell. Therefore, the high level resistance phenotype of the pMB9 series plasmids may be the result of a gene dosage effect due to the multicopy nature of these plasmids.

ACKNOWLEDGMENTS

H.W.B. is an Investigator for the Howard Hughes Medical Institute. This work was supported by grants to H.W.B. from the National Science Foundation (PCM75-10468 A01) and National Institutes of Health (5 R01 CA14026-05). H.L.H., R.L.R. and F.B. were supported by postdoctoral fellowships from the Netherlands Organization for the Advancement of Pure Research (ZWO), the National Cancer Institute, and CONACYT, Mexico, respectively. R.C.T. was supported by a P.H.S. training grant. We would also like to acknowledge David Russel and Alejandra Covarrubias for their discussion and technical assistance.

REFERENCES

(1) S.N. Cohen, A.C.Y. Chang, H.W. Boyer and R. Helling, Proc. Nat. Acad. Sci. USA, 70 (1973) 3240.

(2) J.E. Mertz, R.W. Davis, Proc. Nat. Acad. Sci. USA, 69 (1972) 3370.

(3) J.F. Morrow, S.N. Cohen, A.C.Y. Chang, H.W. Boyer, H.M. Goodman and R.Helling, Proc. Nat. Acad. Sci. USA, 71 (1974) 1743.

(4) R.L. Rodriguez, F. Bolivar, H.M. Goodman, H.W. Boyer and M.C. Betlach, in: Molecular Mechanisms in the Control of Gene Expression, eds. D.P. Nierlich, W.J. Rutter and C.F. Fox (Academic Press, New York, 1976) p. 471.

(5) F. Bolivar, R.L. Rodriguez, M. Betlach and H.W. Boyer, Gene, submitted for publication.

(6) F. Bolivar, R.L. Rodriguez, P.J. Greene, M.C. Betlach, H.L. Heyneker, H.W. Boyer, J.H. Crosa and S. Falkow, Gene, submitted for publication.

(7) S.N. Cohen and A.C.Y. Chang, Proc. Nat. Acad. Sci. USA, 70 (1973) 1293.

(8) R.C. Tait, R.L. Rodriguez and H.W. Boyer, Mol. Gen. Genetics, 151 (1977) 327.

(9) H.I. Adler, W.D. Fisher, A. Cohen, and A.A. Hardigree, Proc. Nat. Acad. Sci. USA, 57 (1966) 321.

(10) D.B. Clewell, J. Bacteriol., 110 (1972) 667.

(11) P. Guerry, D.J. LeBlanc and S. Falkow, J. Bacteriol., 116 (1973) 1064.

(12) D.B. Clewell and D.R. Helinski, J. Bacteriol., 110 (1972) 1135.

(13) R. B. Meagher, R.C. Tait and H.W. Boyer, Cell, 6 (1977) 521.

(14) P.J. Greene, H. Heyneker, M. Betlach, F. Bolivar, R. Rodriguez, A. Covarrubias, I. Fodor and H.W. Boyer, manuscript in preparation.

(15) A. Panet, P.C. Loewen, H.G. Khorana, A.J. Raae, J. R. Lillehaug and K. Kleppe, Biochem., 12 (1973) 5045.

(16) P. Seeburg and H. Schaller, J. Mol. Biol., 92 (1975) 261.

(17) A. Maxam and W. Gilbert, manuscript in preparation.

(18) A.A.Covarrubias, F. Bolivar and H.W. Boyer, manuscript in preparation.

(19) A. Sugino, N.R. Cozarelli, H.L. Heyneker, J. Shine, H.W. Boyer and H.M. Goodman, J. Biol. Chem., submitted for publication.

(20) P.J. Greene, M.S. Poonian, A.L. Nussbaum, L. Tobias, D.E. Garfin, H.W. Boyer, H.M. Goodman, J. Mol. Biol., 99 (1975) 237.

EUCARYOTIC GENOME ORGANIZATION: CLONING AND BEYOND

Bernard Weisblum and Jerry L. Slightom
Pharmacology Dept.,University of Wisconsin, Madison,WI, U.S.A.

The work which we would like to describe falls in the general area of eucaryotic genome organization. Under this heading, two lines of research will be described, one in which recombinant DNA methodology can play a useful role, and one in which it probably cannot because of the large amounts of DNA involved. The eucaryotic organism chosen for these studies is Saccharomyces cerevisiae (Sce) because of its relative simplicity compared to other eucaryotes (10^{10} daltons for Sce vs. 10^{12} daltons for mammals), and the assumption that some of the general features of eucaryotic chromosomal DNA organization are present in a relatively rudimentary form.

REITERATED SEQUENCES IN SCE DNA

Distinguishing properties of reiterated sequences in eucaryotic DNA noted in previous studies have been (1) characteristic density in CsCl density gradients, and (2) accelerated rate of renaturation. Studies of Sce DNA by density gradient centrifugation reveals the presence of two satellite bands- a relatively light, AT-rich "beta" band consisting of mitochondrial DNA, and a relatively dense, GC-rich "gamma" band consisting of the several-hundred-fold reiterated sequences which code for ribosomal RNA. Other types of reiterated sequences in Sce DNA, for example centric DNA, identifiable either as a rapidly renaturing fraction or as a satellite have not yet been found, and the question arises whether other classes of reiterated DNA sequences are present at all. Cloning methodology is ideally suited to answering this question and the answer appears to be yes.

Restriction fragments derived from main band Sce DNA by digestion with endonuclease Eco RI were cloned using the plasmid Col El amp (RSF 2124) as cloning vehicle by the general methods we have described previously (3). A series of 20 plasmid clones obtained in this manner was tested by preparation of cRNA probes using E.coli RNA polymerase, alpha-^{32}P-ATP, and a mixture of 3 complementary unlabelled ribonucleoside triphosphates in the reaction mixture. The labelled cRNA probes were hybridized by the method of Southern (2) to total Sce DNA digested with Eco RI. The hybridization conditions of Denhardt (1) were used (20 mg/liter each of dextran, polyvinylpyrrolidone, and bovine serum albumin in 3XSSC, and 65°C) following which autoradiograms were prepared. The results are shown in Fig.1.

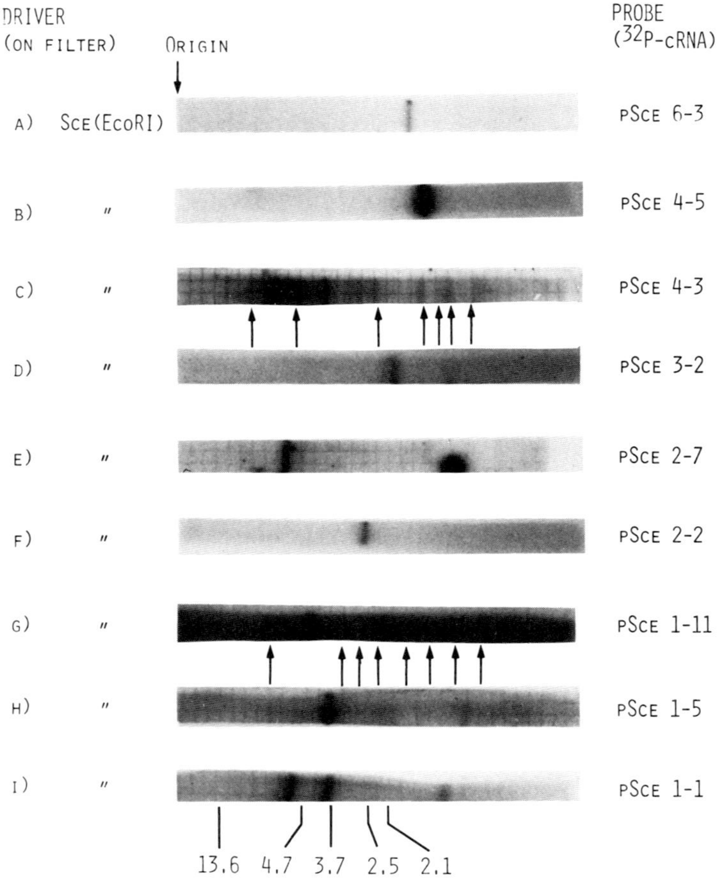

Fig.1. Hybridization of specific cloned fragments to total Sce DNA. Total Sce DNA digested with Eco RI was hybridized by the method of Southern with ^{32}P-cRNA probes prepared from the plasmids listed on the right. Hybridizations c and g contain multiple bands, indicated by arrows and shown more clearly in Fig.2. The numbers at the bottom indicate the position and molecular weight (Mdal) of the 5 fragments obtained by digestion of lambda phage (cb2) DNA with Eco RI.

For the probes tested, single autoradiographic bands were found of which 6 examples are shown in Fig.1a,b,d,e,f, and h. These correspond in electrophoretic mobility to the cloned Sce Eco RI fragment demonstrable as fluorescent bands in the Eco RI digest of the respective recombinant plasmid. In one of the cases shown, Fig.1i, a strong double band corresponding to two fluorescent bands present in Eco RI digests of pSce 1-1 is seen; this is due to the presence of more than one Eco RI fragment in the cloned plasmid. In 2 cases however, Figs.1g and c (pSce 1-11 and pSce 4-3 respectively) additional bands besides those present in the Eco RI digest of the cloned fragments were found.

Digestion of pSce 4-3 with Eco RI followed by electrophoretic analysis of the resultant fragments revealed the presence of 2 fluorescent bands (M.W. 6.2 and 3.75 Mdal) corresponding in mobility to the two autoradiographic bands not designated with arrows. At least 7 additional bands, not present in the Eco RI digest of pSce 4-3 are designated with arrows. Likewise a series of 8 additional bands not present in the Eco RI digest of pSce 1-11 is shown in Fig.1g.

The hybridization profile of the additional bands present in pSce 1-11 and pSce 4-3 is shown more favorably in Fig. 2a and b. For reasons described in detail below, only one, i.e. the smaller of the two Eco RI fragments in pSce 4-3, designated pSce 4-3B was used as template in preparing the probe for this analysis. In Fig.2a,b, the fragment demonstrable in the plasmid clone is designated with an arrow while the homologous bands are clearly visible and range in MW between 2 and 12 Mdal. For comparison, a cRNA probe made from a plasmid containing a single unit of the reiterated rDNA cistrons is shown in Fig.2c. The rDNA probe does not appear to hybridize to any of the components which hybridize with the pSce 1-11 and 4-3 probes. Moreover, there do not appear to be any similarities in the hybridization profiles of pSce 1-11 and pSce 4-3. The membrane filter strips used in this hybridizations were cut sequentially from a nitrocellulose sheet prepared by transfer of the DNA sample fractionated in a slab and applied as a single sample at the origin in a 12 cm trough.

Digestion of pSce 4-3 with Eco RI followed by electrophoretic analysis yielded 3 fluorescent bands, one of which was the vehicle, Col E1 amp (MW 7.3 Md), while the others designated as components A and B respectively of a double passenger correspond in mobility to the two dark bands

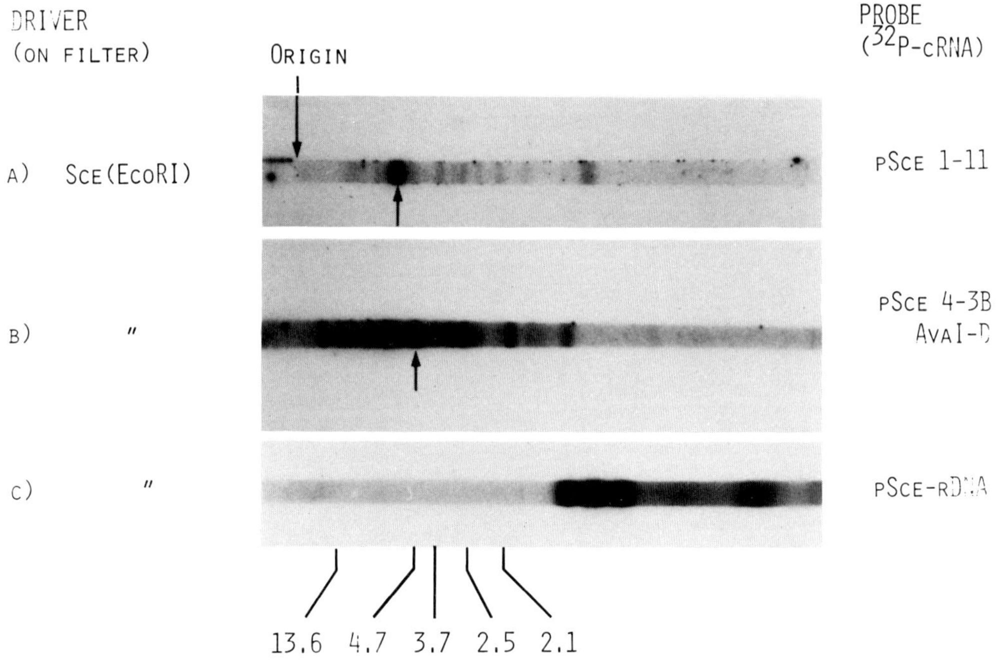

Fig.2. Hybridization of specific cloned fragments to total Sce DNA. Hybridization of ^{32}P-cRNA probes from selected plasmids named on the right to the total profile of Sce DNA digested with Eco RI. Lambda phage DNA standards are as in Fig.1.

shown in Fig.3a; it is not known whether the two fragments are actually contiguous in the Sce genome or whether they represent artifacts of the random ligation of 2 non-contiguous fragments. Following electrophoretic separation, the two passenger fragments were excised from the agarose gel, eluted, and used as templates for preparation of cRNA probes. When the A-specific and B-specific ^{32}P-cRNA probes were used in hybridization experiments, dark bands corresponding to the A- and B-fragments in the Sce profile were seen. For the A-fragment probe hybridization to only a single band was seen (Fig.3b). For the B-fragment probe, however, hybridization to both the B-band as well as to a series of other bands was found (Fig.3c).

In order to obtain a DNA preparation enriched for the reiterated sequences, the B-fragment was subcloned using Col El amp as cloning vehicle yielding a plasmid designated as pSce 4-3B. ^{32}P-cRNA probes were prepared, as above.

pSce 4-3B was digested with AvaI yielding 6 fragments of which 1 was derived completely from Col El amp, 3 were derived completely from Sce, and 2 were "bridge" fragments i.e. which spanned the Eco RI cuts and contained both Col El amp and Sce sequences. Autoradiograms resulting from the hybridization of cRNA probes were prepared from the 6 fragments and maps of the restriction endonuclease cleavage sites based on these results are presented in Fig.3d-i; Fig.4 and Fig.5, respectively.

Of the 6 fragments obtained by AvaI digestion and used as template for cRNA synthesis, fragment AvaI-D showed hybridization to at least 15 DNA bands while fragments AvaI-B,C,E, and F, particularly fragments B and C, showed intense hybridization located primarily over the band corresponding to pSce 4-3B. We conclude from these hybridization experiments that the AvaI-D fragment contains localized reiterated sequences while the AvaI-B,F and C fragments contain predominantly the unique sequences present in pSce 4-3B. Therefore, the multiple bands homologous to the AvaI-D fragment are dispersed in the genome and do not arise by cleavage of a single block of reiterated sequences because fragment D is flanked by F and C.

Finally we show in Fig.6 that both cloned fragments hybridize to main band DNA sequences and not to mitochondrial DNA. The optical density profile and associated hybridization profiles of mitochondrial and total purified main band DNA are shown in Fig.6a and b. pSce 4-3B and pSce 1-11 probes clearly hybridize over the main band fractions as shown in Fig.6c and d.

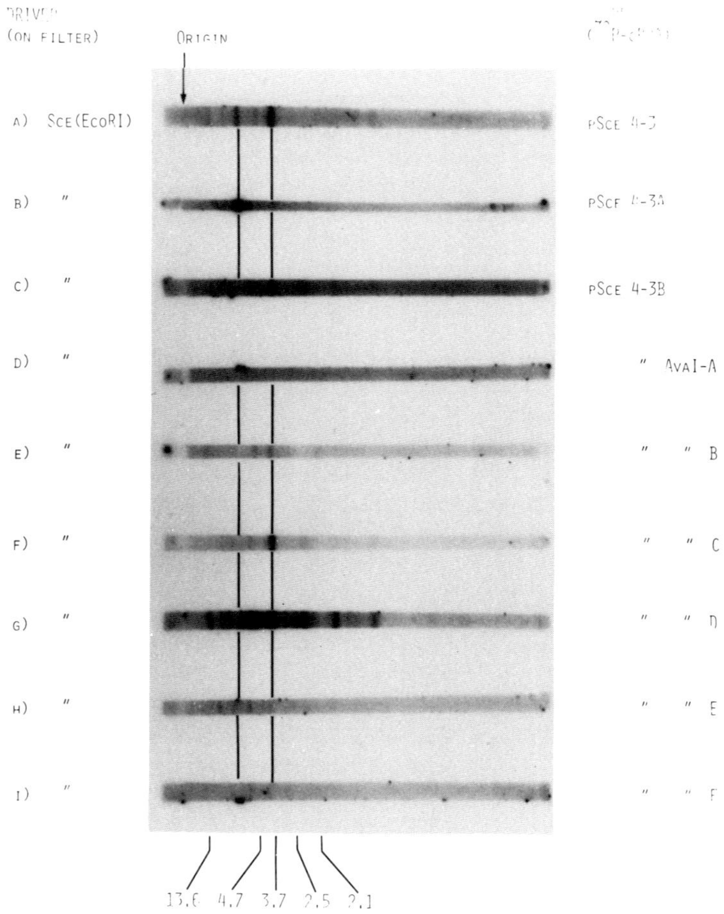

Fig.3. Hybridization of specific DNA fragment probes to total Sce DNA. Hybridization as above with ^{32}P-cRNA probes as indicated to the right. The order of the lettered AvaI fragments is E,A,B,F,D,C,E, where E spans the Eco RI cut at the insertion site of Sce 4-3B into Col E1 amp and the direction E to A proceeds toward the single SmaI site present in Col E1 amp as shown in Fig.5.

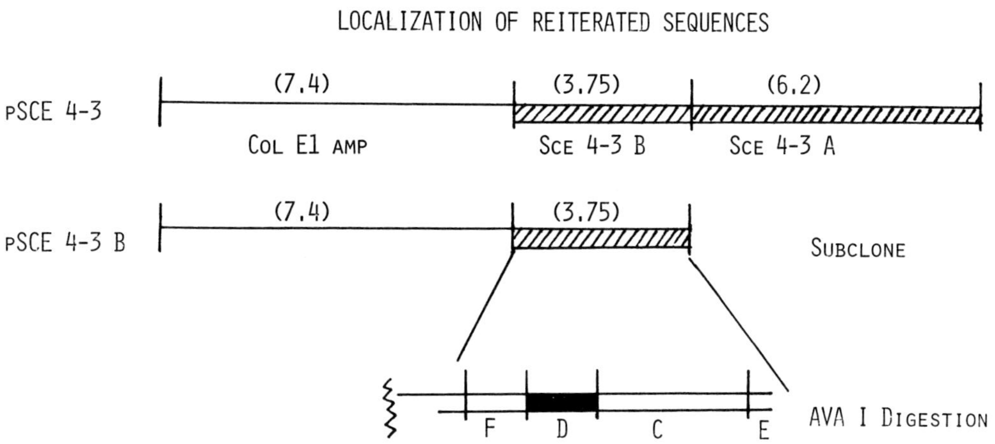

Fig.4. Hybridization of specific DNA fragment probes to total Sce DNA. General schematic maps showing pSce 4-3 and pSce 4-3B. Figures in parenthesis give the molecular weight (Mdal) of the respective fragments, and the shaded segments indicate DNA fragments from which cRNA probes show multiple sites of hybridization to the Eco RI digest of total Sce DNA.

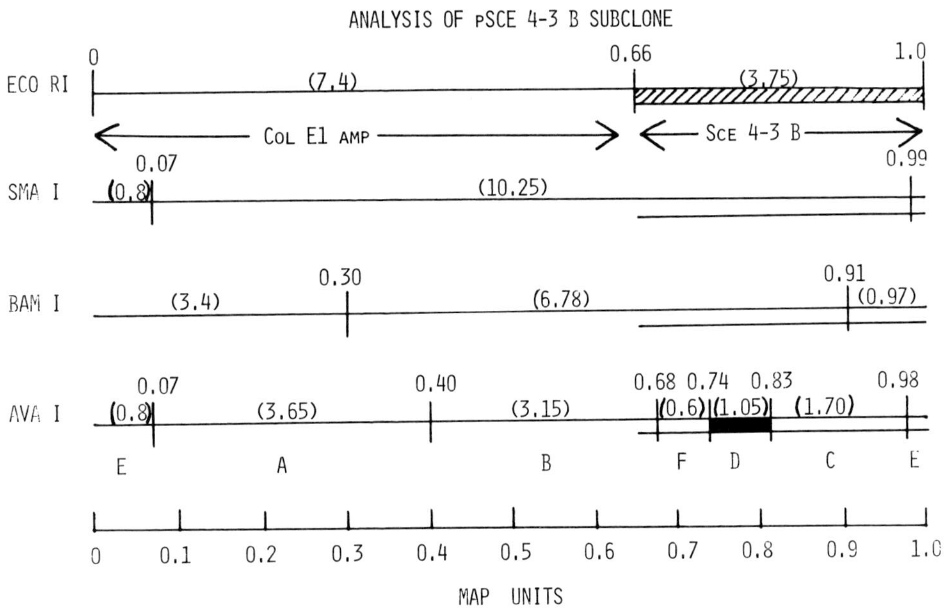

Fig.5. Detailed analysis of pSce 4-3B with restriction endonucleases Eco RI, SmaI, BamI and AvaI. The hatched and filled-in bands denote regions from which transcribed ^{32}P-cRNA probes show multiple sites of hybridization.

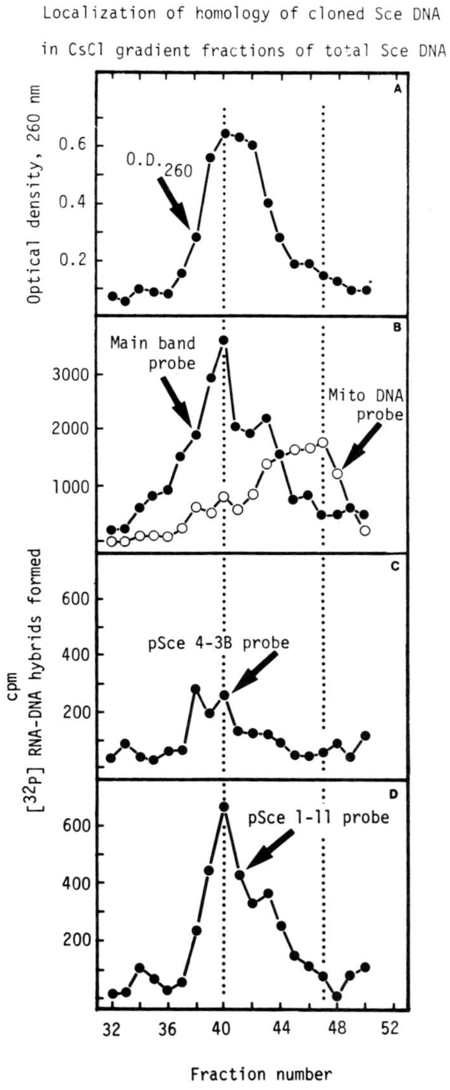

Fig.6. Localization of multiple site homology to CsCl gradient fractions of total Sce DNA. Total Sce DNA was banded in CsCl and successive fractions (32-50, inclusive) transferred to 47 mm Millipore filters for hybridization. Samples, 4 mm in diameter from each gradient fraction on the larger membrane filter were obtained using a paper hole puncher, hybridized with ^{32}P-cRNA probes as indicated, and counted.

BEYOND CLONING

Recombinant DNA techniques have served a useful role in understanding sequence- and organizational aspects of genome fragments less than 10 Md in size- i.e. over the range of fragment sizes which can be cloned using present day technology. Since the MW of the bacterial genome is in the 2000-5000 Md range, this figure would probably constitute an absolute upper limit to the amount of DNA which one could stuff into a "normal" single cell-barring the development of techniques which allow for cultivation of recombinant plasmids in abnormally large multinucleated cells.

Despite these size limitations previous work from our laboratory has shown how information about eucaryotic genome organization in chemical terms can be obtained by cytological methods which employ base-specific fluorescent dyes. The general conclusion from these studies in the light of intensive examination of human chromosomes in other laboratories is that the human genome is organized into domains called "bands" approximately 500 of which are resolvable at the level of light microscopy and have a characteristic base comportion which differ from that of neighboring bands.

We have presented detailed experimental evidence previously (4) from which we deduced that chromosome bands represent regions with characteristic base- and sequence-composition which differ from those which are adjacent. The type of evidence on which this conclusion is based consists of the different patterns of base-specific enhancement or quenching of fluorescence seen in different DNA-dye combinations. In these studies correlation of cytological fluorescence and fluorescence seen with homopolymers or repeating defined repeating polymers was made. The general principles of fluorescent chromosome banding according to DNA base composition and associated fluorescence of defined dye-DNA combinations can be summarized as follows:

1. Fluorescent banding properties of chromosomes can be explained in terms of the fluorometric properties of dye-DNA complexes. In producing fluorescent chromosome bands, the acridines function primarily as reporters rather than as affinity labels. In a sense, the chromosome stains the acridine (as well as vice versa).

2. Acridines which intercalate into DNA can be subdivided into three classes according to the change in their intrinsic fluorescence induced by two DNA test polymers - poly dA·poly dT and poly dG·poly dC; these are quenching, quenching, respectively (e.g. 9-aminoacridine); enhancement, quenching, respectively (e.g. quinacrine and proflavine);

enhancement, enhancement, respectively (e.g. acridine orange and coriphosphine O).

3. There exists a direct correlation between the fluorometric properties of dye-DNA complexes and fluorescent patterns observed in cytologic preparations. Thus 9-aminoacridine serves no useful function as a cytologic stain; fluorescence due to acridine orange is generalized and non-discriminating; cytologic fluorescence due to quinacrine is localized to regions which contain a high density of uninterrupted AT base pairs.

4. The two main determinants of quinacrine fluorescence are enhancement due to uninterrupted stretches of AT base pairs and quenching due to GC base pairs (actually G-alone is sufficient).

5. Treatment of human metaphase chromosomes with acridine orange after an initial pretreatment with 0.1 M NaCl at neutral pH and 85°C can be used to distinguish both AT-rich and GC-rich regions. Under these conditions, presumed to result in partial denaturation of chromosomal DNA, AT-rich regions are preferentially denatured and fluoresce red, while GC-rich regions remain relatively less denatured and exhibit a yellow to green fluorescence. Results obtained by this method are consistent with results obtained with quinacrine and with the bibenzimidazole dye Hoechst 33258 described below.

6. Three qualities of cytologic fluorescence demonstrated with quinacrine are: (a) brilliant fluorescence (e.g. the Y-chromosome (long arm) in man, and in D. melanogaster); (b) bright fluorescent bands not easily seen with the eye but demonstratable with high contrast photographic techniques (e.g. Q bands in human chromosomes); (c) relatively dark or non-fluorescing bands (e.g. distal tip of 1p in man; centric regions of murine and bovine chromosomes).

7. The three qualities of cytologic fluorescence with quinacrine can be accounted for solely by the base-composition of the DNA present and the degree of interspersion or periodicity of the GC base pairs. According to the reporter model: (a) For two polynucleotides with the same base composition, the fluorescence of intercalated quinacrine will show greater quenching in the case of the polynucleotide in which the GC base pairs are distributed with greater periodicity (cf. dA·dT + dG·dC, 1:1, 50% GC, which enhances fluorescence) vs. E. coli DNA (also 50% GC, but which quenches fluorescence); (b) For two polynucleotides with different base compositions, the

polynucleotide with less GC (e.g. C. perfringens DNA 32% GC) will actually quench more than a polynucleotide with more GC (e.g. dA·dT + dG·dC, 1:1, 50% GC) if, as in the case of the former, the AT and GC base pairs are more evenly interspersed.

8. Centric heterochromatin being highly repetitive exhibits non- or negative fluorescence in quinacrine-treated preparations over a wide range of base compositions (35 to 55% GC). The degree of AT-richness of the centric heterochromatin relative to euchromatic chromosome regions can be established with aid of Hoechst 33258, a compound whose fluorescence is enhanced strongly by AT base pairs and enhanced weakly (but definitely not quenched as in the case of quinacrine) by GC base pairs. Hoechst 33258 serves as a touchstone to discriminate between the "AT-rich" (35% GC) murine centric heterochromatin and the "GC-rich" (55% GC) bovine centric heterochromatin, giving bright fluorescence in the case of the former and dull fluorescence in the case of the latter.

From a study of bright and dark bands seen in quinacrine-treated chromosomes it is possible to infer that, with the exception of highly repetitive centric DNA, bright telomeric bands contain DNA relatively richer in AT than the dark bands. An instructive example is that of human chromosome number one. The distal segment of 1p is relatively quinacrine dark, while a large segment of 1p proximal to the centric region is relatively bright. The inference to be drawn from such observations is that the distal tip of 1p is relatively GC-rich while more proximally the DNA is relatively richer in AT. Further data which favor this interpretation come from the AO/85 method which indicates that the distal tip of 1p is relatively difficult to denature and more directly from studies with anti-cytidine antibodies which show preferential affinity for the distal segment of 1p and in general produces a reversed banding pattern. By this type of analysis, the distal tips of most of the human chromosomes are enriched for DNA relatively rich in GC.

In terms of unique genes and interspersed repetitive sequences, at least three models can be proposed to account for the observed cytological patterns:

1. Quinacrine-dark bands contain unique gene sequences with higher GC than bright bands. This type of explanation would be consistent with the observation that bacterial DNA's (consisting mostly of structural gene sequences) can range in base composition from 32% GC (Clostridium perfringens) to 72% GC (Micrococcus luteus).

2. Unique gene sequences in both quinacrine-bright and dark bands have a similar average base composition; however, GC-rich interspersed repetitive sequences are responsible for dark bands while AT-rich interspersed repetitive sequences are responsible for bright bands.

3. Interspersed repetitive sequences are generally enriched for AT base pairs while unique gene sequences are generally enriched in GC base pairs. Dark bands, however, contain a relatively higher concentration of unique gene sequences relative to interspersed repetitive sequences than the bright bands.

4. Some combination of the above.

If we assume that the bands seen by fluorescence microscopy number between 100 and 1000, we can estimate that each band contains on the average between 10^9 and 10^{10} daltons of DNA. Thus from direct observation of metaphase chromosomes we can actually see a pattern of organization of the human caryotype into regions with characteristically different average base composition. This differentiation into bright and dark telomeric bands also serves to distinguish late and early replicating sequences in the genome. While centric regions replicate late regardless whether they are AT- or GC-rich.

The relationship between models of genome organization responsible for observed banding patterns on one hand, and the relationship between the various classes of unique and reiterated sequences seen in cloned fragments of eucaryotic DNA on the other, is part of the task before us.

References

1. Denhardt, D. Biochem. Biophys. Res. Commun. 23; 641 (1966).
2. Southern, E.M. J. Mol. Biol. 98; 503 (1975).
3. Tanaka, T. and Weisblum, B. J. Bacteriol. 121; 354 (1975).
4. Weisblum, B. Cold Spring Harbor Symp. Quant. Biol. 38; 441 (1973).

CONSTRUCTION AND USE OF SAFER BACTERIAL HOST STRAINS FOR RECOMBINANT DNA RESEARCH

ROY CURTISS III, MATSUHISA INOUE, DENNIS PEREIRA,
J. CHARLES HSU, LAURA ALEXANDER and LYNN ROCK
Department of Microbiology
Institute of Dental Research
Cancer Research and Training Center
University of Alabama in Birmingham
Birmingham, Alabama 35294 U.S.A.

Abstract: We have designed, constructed and tested a strain of Escherichia coli K-12 as a safer host for use with plasmid cloning vectors during recombinant DNA research. This strain, designated χ1776, is in many respects quite different from most E. coli strains that scientists are accustomed to using. Because of this, it has been necessary to devise new procedures to grow, transform and recover amplified plasmid DNA from this strain. Due to some of χ1776's unique attributes, it has been possible to devise a method to lyse it without use of detergents. This method provides higher yields of plasmid DNA than can be recovered from conventional strains of E. coli by standard methods. We also report our progress in constructing improved derivatives of χ1776 that possess additional safety features and also on the design and construction of safer host strains for use with various λ vectors.

INTRODUCTION

For the past two years, our laboratory group has been designing, constructing and testing safer, more useful strains of Escherichia coli K-12 for recombinant DNA molecule research. Our goal has been to construct strains that possess mutations that: (i) increase usefulness for research, (ii) preclude synthesis of cell wall in non-laboratory-controlled environments, (iii) lead to degradation of DNA in non-laboratory-controlled environments, (iv) preclude colonization of and survival in the intestinal tract of warm-blooded animals, (v) contribute to cell death in non-laboratory-controlled environments, (vi) permit cloning vector replication to be dependent on host and/or laboratory-controlled conditions, (vii) preclude or minimize transmission of recombinant DNA to other bacteria, and (viii) permit monitoring and prevention of contamination during transformation or transfection. We have during the course of these studies learned a considerable amount of information about those mutations that contribute to the utility and/or safety of strains which possess them (1).

PROPERTIES OF χ1776

E. coli K-12 strain χ1776 was derived in 13 steps from the minicell-producing strain χ1276 (1,2). χ1776 has the genotype F^- tonA53 dapD8 minA1 supE42 Δ40[gal-uvrB] λ^- minB2 rfb-2 nalA25 oms-2 thyA57* metC65 oms-1 Δ29[bioH-asd] cycB2 cycA1 hsdR2. χ1776 has been certified by the National Institutes of Health as an EK2 host-vector system in conjunction with the pSC101 and pCR1 non-conjugative plasmid cloning vectors.

χ1776 has a number of useful features. It is restrictionless (due to the hsdR2 allele), thus facilitating the introduction of foreign DNA. It is easy to transform with plasmid DNA (due to dapD8, Δ29[bioH-asd] and Δ40[gal-uvrB] mutations), as will be briefly mentioned later in this report and is more fully described elsewhere (3). It is easy to lyse even without detergents in order to recover recombinant plasmid DNA (due to the dapD8, Δ29[bioH-asd], Δ40[gal-uvrB], rfb-2, oms-1 and oms-2 mutations). The procedures for accomplishing this as well as for the amplification in copy number of ColE1-derived plasmid vectors will be described below. Lastly, χ1776 produces minicells (due to minA1 and minB2 mutations) which should facilitate studies on recombinant DNA molecule expression in the absence of background noise caused by chromosomal gene expression (4,5,6).

In terms of safety features, χ1776 has an obligate requirement for diaminopimelic acid (DAP; due to dapD8 and Δ29[bioH-asd] mutations) and cannot synthesize the mucopolysaccharide colanic acid (due to the Δ40[gal-uvrB] mutation). It thus undergoes DAP-less death in all environments where DAP is limiting but sufficient nutrients exist to support cellular metabolism and growth. It requires thymine or thymidine (due to thyA57* mutation) and undergoes thymineless death with degradation of DNA when thymine and thymidine are absent from the environment but when sufficient nutrients are present to sustain metabolic activity. χ1776 is extremely sensitive to bile (due to rfb-2, oms-1 and oms-2 mutations) and thus is unable to survive passage through the intestinal tract of rats. These same three mutations confer upon χ1776 extreme sensitivity to detergents, antibiotics, drugs and chemicals and thus cause it to be less likely to survive in environments in which these compounds might be encountered, such as in sewers. Due to the Δ40[gal-uvrB] mutation, χ1776 is unable to carry out either dark or photo repair of UV-induced damage and is thus several orders of magnitude more sensitive to sunlight than wild-type strains of *E. coli*. χ1776 is resistant to many transducing phages and is conjugation deficient for inheritance of many different types of conjugative plasmids due to the presence of the tonA53, Δ29[bioH-asd], Δ40[gal-uvrB], rfb-2, oms-1 and oms-2 mutations (7). Although conjugational transmission of cloning vector DNA in triparental matings is detectable under optimal permissive laboratory conditions, neither transductional nor conjugational transmission of cloning vector DNA is detectable under a variety of nonpermissive conditions that might be analogous to those encountered in nature (7). χ1776 is resistant to nalidixic acid (nalA25), cycloserine (cycA1, cycB2) and trimethoprim (thyA57*). These drugs can therefore be added to media to permit monitoring of the strain and to preclude transformation of contaminants during transformation.

Since χ1776 has a rather fragile cell wall and exhibits extreme sensitivity to detergents, it is essential that all glassware and plasticware used in the cultivation of this strain be extremely clean and well rinsed so as to be free from all residual detergents. χ1776 grows with a generation time of about 50 min in either L broth or Penassay broth when supplemented with 100 μg DAP/ml and 4 μg thymidine/ml and reaches final densities of 8-10 x 10^8 cells/ml at stationary phase. χ1776 is particularly sensitive to mechanical shear due to its fragile cell wall and somewhat elongated cell shape. It is thus essential to avoid use of vortex mixers when suspending cells pelleted by centrifugation. We find that gentle agitation by swirling and shaking back and forth for a period of 1-2 min adequately suspends cells with maintenance of 100% viability. To illustrate, if a χ1776 culture is sedimented and then the pellet suspended by vortex mixing for 60 sec, between 90 to 99.9% of the cells die with many of them lysing. We therefore recommend to those individuals who have had difficulty in the growth and maintenance of χ1776 to exhibit a little more tender loving care. χ1776 will reward them for their efforts.

RECOVERY OF PLASMID DNA FROM χ1776

During transformation of χ1776 with plasmid DNA purified following detergent-facilitated lysis of spheroplasts, we soon noticed that residual amounts of detergent in the plasmid DNA were extremely inhibitory (3). Because of this and the known fragility of χ1776's cell wall, we decided to see if it would be possible to devise a method for the detergent-independent lysis of plasmid-containing derivatives of χ1776. We were soon able to show that lysozyme-generated spheroplasts could be readily lysed by dilution with ice-cold water followed by a heat shock and then further incubation at 0C. The optimal method so far devised for the recovery of plasmid DNA from χ1776 derivatives is as follows: Cells grown with aeration in a variety of media are sedimented, washed twice and gently suspended in TEN buffer (20 mM Tris, 20 mM EDTA, 0.8% NaCl, pH 8.0). Lysozyme is added to 200 μg/ml and the suspension incubated at room temperature for 20 min. The spheroplast mixture is then placed on ice and an equal volume of ice-cold water added. After about 5 min the mixture is placed at 65 to 68C for 5 min. The tube containing the mixture is then placed on ice and held overnight before further processing. We have noted in several experiments that holding the lysate for two days at 0C may be better in terms of plasmid recovery if the preparation is to be centrifuged in order to remove chromosomal DNA. The chromosomal DNA is sedimented in an SW41 rotor in a Beckman preparative ultracentrifuge at 30,000 rpm for 25 min at 3-4C. This cleared lysate can be used directly for the transformation of bacterial strains or the plasmid DNA can be further purified by ethidium bromide-CsCl gradient centrifugation (8). Table 1 presents data on the yield of plasmid DNA from derivatives of χ1776 containing the plasmids pSC101, R6K and R100*drd1*. As can be seen from the data, slightly higher yields of covalently-closed circular (CCC) plasmid DNA molecules can be isolated when the lysozyme-generated spheroplasts are lysed without the use of either 0.1% Brij plus 0.1% deoxycholate or 0.1% sarkosyl. We have also noted that the average number of copies of CCC plasmid DNA per chromosome DNA equivalent is somewhat higher for the pSC101 and R100*drd1* plasmids when spheroplasts are lysed without use of detergents than for values reported by other investigators using detergent-facilitated lysis of spheroplasts. Of course, these differences may be dependent upon the strain as well as the method.

TABLE 1

Yield of plasmid DNA from χ1776 derivatives*

Plasmid	Detergent for cell lysis	Percent CCC DNA/ chromosome DNA Exp. 1	Exp. 2	Exp. 3	Copies/ chromosome
pSC101	none	1.8	1.7	1.7†	4-6
	Brij + DOC	1.6	-	-	
	Sarkosyl	1.4	1.4	-	
R6K	none	13.7	14.3		11-15
	Brij + DOC	9.8	10.4		
	Sarkosyl	11.1	11.6		
R100*drd1*	none	5.4	5.8		2-3
	Brij + DOC	4.8	5.5		
	Sarkosyl	5.3	6.0		

* Spheroplasts were lysed as described in text or by addition of 0.1% Brij-58 plus 0.1% deoxycholate or by addition of 0.1% sarkosyl. Whole cell lysates were fractionated on EtBr-CsCl gradients.

†If held lysate for one week at 0C, yield increased to 3.0%.

Table 2 presents data on plasmid yields from cleared lysates and whole cells of χ1776 derivatives. The data reveal a substantial enrichment in the percent CCC plasmid DNA/chromosome DNA for the material from the cleared lysates as opposed to that from the whole cell lysates. It should be noted that in one experiment the percent recovery of pMB9 plasmid DNA approached 100% when the cleared lysate was held at 0C for two days prior to the clearing centrifugation. We have observed this behavior on several occasions and are now determining the optimal period of time for holding the lysate at 0C prior to the clearing spin so as to maximize the recovery and purity of plasmid DNA. If we can obtain cleared lysates in which over 80% of the DNA is plasmid DNA, the need for

TABLE 2

Plasmid yields from cleared lysates and whole cells of χ1776 derivatives

Plasmid	Lysate*	Percent recovery		Percent CCC DNA/
		Chromosome	Plasmid	chromosome DNA
pSC101	Whole cells	100	100	1.7
	Cleared[†]	1.1	40	61.5
pMB9	Whole cells	100	100	6.4
	Cleared[†]	3.2	28[‡]	103
R6K	Whole cells	100	100	11.4
	Cleared[†]	5.6	43	86

* Lysozyme-heat-ice shock method (no detergents).

† Lysate held overnight at 0C and then centrifuged at 30,000 rpm in SW41 rotor for 25 min at 3-4C.

‡ If lysate held at 0C for two days prior to clearing centrifugation, recovered 97% of plasmid DNA.

further plasmid purification by ethidium bromide-CsCl gradient centrifugation followed by extensive dialysis will be eliminated.

In order to maximize the recovery of ColE1-derived plasmid cloning vectors, we next turned our attention to developing optimal conditions for the amplification of plasmid copy number by incubation of cells in the presence of chloramphenicol. Most procedures for the amplification of ColE1 plasmid DNA in more "normal" strains of E. coli specify the incubation of cultures overnight in the presence of 100 to 200 μg chloramphenicol/ml (9). In our initial experiments, we evaluated viable cell titers of χ2042, a χ1776 derivative containing the pMB9 cloning vector, following overnight incubation in the presence of different concentrations of chloramphenicol. As the data in Table 3 reveal, there is a significant amount of cell death even when using a low concentration of chloramphenicol such as 12.5 μg/ml. In these same experiments we also measured the minimal inhibitory concentrations for tetracycline and found that these were lower in cultures treated with concentrations of 50, 100 or 200 μg chloramphenicol/ml than they were for cells incubated in the absense of chloramphenicol.

TABLE 3

Survival of χ2042 after 22 hours incubation in various concentrations of chloramphenicol (Cm)*

Concentration of Cm (μg/ml)	Viable titer	Percent survival
0	4.7×10^8	100
12.5	1.4×10^8	30
25	1.3×10^8	28
50	4.0×10^7	8.5
100	2.0×10^7	4.3
200	8.0×10^6	1.7

* Cells grown in L broth with 100 μg DAP/ml and 4 μg Thd/ml.

We next conducted a series of experiments in which χ2042 cells were prelabeled with [^{3}H]-thimidine, washed, and suspended in prewarmed fresh media with various concentrations of chloramphenicol and [^{14}C]-thymidine so that we could follow the fate of [^{3}H]-labeled chromosomal DNA as well as [^{14}C] incorporation into plasmid DNA. In these experiments, we found that incubation of cells in the presence of 100 μg Cm/ml led to detectable losses in chromosomal DNA within the first several hours of incubation. Even when the concentration of chloramphenicol was reduced to 12.5 μg/ml, cell death with loss of prelabeled counts became readily observable following six hours of incubation. Based on studies of these types, we found the following optimal conditions for maximal recovery of ColE1-derived plasmid vectors. A standing overnight culture is diluted 1:10 into the same prewarmed growth medium and aerated by shaking until reaching an A_{600} of about 0.45 ($\sim 3 \times 10^8$ cells/ml). Chloramphenicol is then added to 12.5 μg/ml and the culture is aerated for an additional five hours. The cells are then harvested and suspended in TEN buffer for spheroplasting and isolation of plasmid DNA. It should be reiterated that the suspension of cells should be by gentle swirling and agitation. The spheroplasts are lysed by the ice water-heat-cold treatment with no detergent. Table 4 presents data on the amplification of pMB9 by chloramphenicol in χ1776.

TABLE 4

Amplification of pMB9 with chloramphenicol (Cm)*

Cm concentration (μg/ml)	Percent CCC DNA/ chromosome DNA	Plasmid copies/ chromosome
0	6.4	49
12.5	60	463
100	67	516
0	7.2	60
5.0	34	323
12.5	44	418

* Cm treatment for 5 hours. Cells lysed by lysozyme-heat-ice shock method. Lysates were fractionated on EtBr-CsCl gradients.

Although treatment of the culture with 100 μg chloramphenicol/ml gave the highest plasmid copy number per chromosomal DNA equivalent, the total yield of plasmid DNA was very much less than that recovered when the culture was treated with 12.5 μg chloramphenicol/ml.

IMPROVED METHODS FOR TRANSFORMATION OF χ1776

By careful studies of all possible variables for growth of cells, induction of competence, uptake of plasmid DNA and recovery of transformants, it has been possible to improve transformation frequencies, not only for χ1776 but for other strains of *E. coli* (3). In terms of transformation of χ1776, we have already mentioned the necessity of using glassware and plasticware that is completely free from detergents. Indeed, we routinely use plasticware and Pyrex test tubes that have not been placed in general laboratory circulation and that are cleaned without detergents or brushing. We also use DNA that either has been obtained by using cell lysis methods not employing detergents or has been dialyzed extensively for several days. Cultures of χ1776 are grown to mid-log phase in L broth with aeration and are harvested when they reach a density of 2-3 x 10^8 cells/ml. The cells are sedimented at 8000 rpm in a Sorvall SS34 rotor at room temperature for 10 min. The cells are gently suspended in 10 mM NaCl and then resedimented and suspended in 75 mM $CaCl_2$ containing 0.8% NaCl and 10 mM Tris (pH 8.0). After 20 min at room temperature, the cells are resedimented and suspended in one-tenth the original culture volume in the same $CaCl_2$-NaCl-Tris buffer. Two minutes prior to addition of DNA, 200 μl samples in Pyrex test tubes are placed on ice. Following the addition of 100 μl of plasmid DNA (0.1 μg/ml in 20 mM Tris-0.8% NaCl, pH 8.4 at 4C), the cells are held on ice for 20 min and then heat shocked at 42C for exactly 1.0 min. The mixture is then placed at room temperature for 10 min and samples are plated directly on freshly prepared EMB agar containing 12.5 μg tetracycline/ml, 25 μg nalidixic acid/ml and/or 15 μg cycloserine/ml. These selective plates should be prepared the same day and the plates should not be spread to dryness. The use of overly dry plates and/or spreading to dryness can reduce transformant yield as much as 1000-fold. If the plasmid cloning vector expresses resistance to ampicillin or one of the aminoglycoside antibiotics, the transformation mixture should be diluted 1:10 into L broth and incubated for 90 to 120 min prior to plating. It should be noted that the transformant yield only increases two- to three-fold when the cell density is increased from 10^7/ml to 10^9/ml. We believe this is due to the fact that the occurrence of some cell lysis leads to the release of chromosomal DNA which is known to compete for uptake of plasmid DNA and to reduce the yield of transformants inheriting any of the plasmid vectors. When using plasmid DNA at a concentration of 0.1μg/ml, we are able to routinely achieve transformation frequencies of between 10^{-2} to slightly in excess of 10^{-1}/μg of plasmid DNA/surviving cell.

CONSTRUCTION OF SAFER HOSTS FOR λ VECTORS

It became apparent in June 1975 that many of the safer λ vectors being constructed for use in recombinant

DNA research were still able to form lysogens at frequencies in excess of 10^{-8} in the propagating host. It was therefore deemed advisable to commence to genetically alter bacterial strains that would provide useful, safer hosts for use with these λ vectors. In addition to the general goals enumerated in the Introduction of this manuscript, we established four additional goals in constructing these strains. First, we wanted the strains to permit yields of λ in excess of 10^{10} infectious particles/ml. Second, we wanted the strains to be highly transfectable so that we could obtain transfection frequencies of about 10^{-4}/λ genome equivalent in the absence of any need to spheroplast the cells. Third, we wanted the strains to possess genetic markers that would facilitate the efficient monitoring of lysogen formation by the λ vectors and fourth, we wanted to decrease any unnecessary genome homology between constructed λ chimeras and the host chromosome.

We started with strain χ1038 (F^- lacY1 supE44 galK2 galT22 λ^- metB1 hsdS3) which is a single colony isolate of Arber's strain 803 (10,11). Although we have not quite completed the construction of the final two strains, we have in 13 genetic steps constructed χ1960 which has the genotype F^- Δdap-21 ΔlacZ39 supE44 Δ47[gal-uvrB] λ^- ΔtrpE63 nalA29 ΔthyA57 endA1 hsdS3 serB31. It should be noted that χ1960 has retained maximum sensitivity to λ, ϕ80 and 434, which should enable use of a final derivative of this strain for λdv cloning vectors. All the mutations so far introduced into this strain which are essential for the strain's reduced ability to survive in nature (Δdap-21, Δ47[gal-uvrB] and ΔthyA57) are unable to revert spontaneously or following mutagen treatment. The presence of the ΔlacZ39 mutation should facilitate monitoring for lysogen formation of λ vectors derived from λplac5 and the Δ47[gal-uvrB] mutation should facilitate monitoring lysogen formation with phage vectors that possess either gal or bio chromosomal inserts. The Δ47[gal-uvrB] mutation, of course, deletes the normal integration site for λ prophage and blocks both dark and photo repair of UV-induced damage. This mutation also increases transfectability with λ DNA about 30-fold over that observed with a gal^+ strain (Gurnam Gill, personal communication). The endA1 mutation does not alter transfectability but does increase transformability with plasmid DNA 3- to 10-fold (1).

In futher modifications of χ1960 we intend to first

introduce a ΔdeoA mutation by cotransduction to $serB^+$. This will provide a strain that will have an obligate requirement for thymidine which should increase the likelihood of thymineless death in natural environments where thymine seems to be more prevalent than thymidine. Our next step will be an attempt to remove the supE44 mutation so as to yield a strain free of amber suppressors. Such a strain will be useful in testing λ vectors containing cloned DNA for the maintenance of the amber suppressor mutations. In a separate strain we will introduce the strong nonsense suppressor tyrT58 (supF58) by cotransduction with $trpE^+$. The presence of this suppressor not only strongly suppresses amber mutations present in many λ cloning vectors but will also permit the strain to propagate λ vectors having amber mutations in the S gene.

Our ultimate goal in constructing safer hosts for λ vectors is to develop a host with a very much diminished ability to survive in non-laboratory-controlled environments so that (i) a suppressor-free derivative could be used as a host for λ lysogens and (ii) a suppressor-containing derivative could be used for λdv and other plasmid cloning vectors.

CONSTRUCTION OF SAFER, MORE USEFUL DERIVATIVES OF χ1776

In the year since the isolation of χ1776 in January 1975, we have learned a good deal about its properties and have found certain attributes which can stand improvement. For example, when χ1776 was constructed it was thought to possess the ΔthyA57 mutation which had never been observed to revert. Nevertheless, the thyA57* mutation in χ1776 does revert at a low frequency ($\sim 10^{-9}$) and the strain can also accumulate deo mutations that permit growth of the strain on low levels of thymine. Secondly, χ1776 can mutate to become resistant to bile and detergents at a low frequency ($\sim 10^{-9}$). Lastly, χ1776 is not as defective as might be desired with regard to its ability to inherit F and I group conjugative plasmids which are prevalent in many E. coli strains encountered in nature. For these reasons, we have commenced to modify χ1776 so as to make it more nearly perfect and to accomplish several other objectives. So far, we have constructed in five steps χ2070 (F^- tonA53 dapD8 minA1 supE42 Δ40[gal-uvrB] λ^- minB2 rfb-2 nalA25 ΔthyA57 endA1 metC65 oms-1 Δ29[bioH-asd] cycB2 cycA1 hsdR2 serB31). In introducing the ΔthyA57 mutation, which does not revert to thyA$^+$, we had to remove the oms-2 mutation which contributed some of the bile sensitivity to χ1776. χ2070 is therefore not as bile and detergent sensitive as its ancestor. The endA1 mutation was introduced since as previously noted it increases transformability with plasmid DNA 3- to 10-fold (1). The next steps will be to introduce the ΔdeoA deletion by cotransduction with serB$^+$. We then intend to introduce a ΔgalU mutation which should contribute to bile and detergent sensitivity and also provide a second block against the synthesis of the mucopolysaccharide capsular material colanic acid. The ΔaraC766 and ΔtrpE63 mutations will be introduced into the strain to facilitate the use of plasmid cloning vectors wherein the selectable markers are araC and trpE. The final step will be to introduce lpc and omp mutations to make the strain completely bile and detergent sensitive and unable to revert to resistance to these compounds. The lpc and omp mutations will also greatly diminish the ability of the strain to act as a recipient for the F and I group conjugative plasmids.

CONCLUDING DISCUSSION

It has become evident to us that the design and construction of derivatives of *E. coli* K-12 as safer hosts for plasmid and phage cloning vectors has resulted in strains which no longer possess attributes customarily associated with members of the *E. coli* species. Although this initially caused great difficulty on the part of many who commenced to use these strains and was a frustration to us, we have subsequently learned that these strains have many attributes that make them more useful for plasmid research in general and recombinant DNA research in particular than available standard strains of *E. coli*.

ACKNOWLEDGMENT

Research supported by funds provided by the National Institutes of Health (DE-02670, AI-62508, CA-13148, AI-07041, T32 GM07164) and by the National Science Foundation (GB-37546).

REFERENCES

(1) R. Curtiss III, D. A. Pereira, J. C. Hsu, S. C. Hull, J. E. Clark, L. J. Maturin, Sr., R. Goldschmidt, R.

Moody, M. Inoue and L. Alexander. Tenth Miles International Symposium on Impact of Recombinant Molecules on Science and Society (1976).

(2) R. Curtiss III, D. Pereira, J. E. Clark, J. C. Hsu, R. Goldschmidt, S. I. Hull, R. Moody, L. Maturin and M. Inoue. In preparation.

(3) M. Inoue and R. Curtiss III. In preparation.

(4) A. C. Frazer and R. Curtiss III. Curr. Topics Microbiol. Immunol. 69(1975)1.

(5) F. Morrow, S. N. Cohen, A.C.Y. Chang, H. W. Boyer, M. H. Goodman and R. B. Helling. Proc. Natl. Acad. Sci. U.S.A. 71(1974)1743.

(6) A.C.Y. Chang, R. A. Lansman, D. A. Clayton and S. N. Cohen. Cell. 6(1975)231.

(7) J. E. Clark-Curtiss, J. R. Moody, Jr., L. Alexander, J. C. Hsu and R. Curtiss III. In preparation.

(8) R. Radloff, W. Bauer and J. Vinograd. Proc. Natl. Acad. Sci. U.S.A. 57(1967)1514.

(9) D. B. Clewell. J. Bacteriol. 110(1972)667.

(10) W. B. Wood. J. Mol. Biol. 16(1966)118.

(11) G. Kellenberger, N. Symonds and W. Arber. Vererbungsl. 98(1966)247.

DISCUSSION

K. SAKAGUCHI: What is the maximal growth of your strain?

R. CURTISS: The maximal densities attainable for χ 1776 in late-log to early stationery phase are pretty close to 1 x 10^9 viable cells per ml. The generation time in rich media is about 50 minutes. So one can actually let the bacteria grow to somewhat higher densities than we did in these experiments before adding chloramphenical to amplify plasmid copy numbers. I see no reason why one couldn't go to higher densities. I think one can make a number of modifications with equally good, or, I would hope better, results. The only reason we added chloramphical to cells at 3 x 10^8/ml is because they reach that density in about 4 hours and after you go through 5 hours of amplification you still have another few hours for cell lysis and recovery of plasmid DNA; it makes for a nice 12 hour day, rather than stretch it out any longer than that.

C. WEISSMANN: How many micrograms of plasmid DNA can you recover from the χ 1776 and also, how many transformants per microgram of plasma DNA do you get?

R. CURTISS: Following chloramphenical amplification of pMB9 under the conditions used, we are getting between 1 to 2 micrograms of plasmid DNA per ml of culture. I imagine that by increasing the density of the culture the yield can be further increased. We get almost 106 transformants per microgram of plasmid DNA depending on which plasmid and its molecular weight. For pSC101, pMB9 and pCR1 we are obtaining mean frequencies of about 5 x 10^{-6} per plasmid molecule for several different recipients including χ 1776. We are still improving the transformant yield; I don't know what the limit is. About 60 to 80% of the cells die during the transformation process. If we can eliminate that amount of cell death and lysis, we may be able to reduce the amount of chromosomal DNA released which competes for uptake of plasmid DNA and thereby get an absolute increase of at least another five fold. With phage transfection, we are getting 1 tranfectant per 10^4 λ, T1 or T7 genome equivalents. That is still higher than we are getting with plasmid transformation.

C. WEISSMANN: In your published procedure in the NARSM bulletin, you do not incubate your cells prior to plating. Don't you have to give some time to express the resistance.

R. CURTISS: If the selectable marker is tetracycline resistance, we observe the same yield of transformants when plating immediately or after allowing a period of growth. This

I agree, is an unexpected observation. There are at least three possible contributing factors to explain these results. First, tetracycline is bacteriostatic rather than bacteriocidal; thus even if transformants could slowly express resistance on the selective plates they would not be killed prior to expression and would eventually grow up into colonies. Second, tetracycline resistance in pMB9 at least is constitutive as mentioned by Herb Boyer this morning. Third, the transformed cells may be stabilized by being placed on an agar surface. Following the transformation procedure, the cells are quite fragile and we may thus have some extra cell death when they are diluted into growth media which would offset any increase in transformant yield caused by allowing for expression in the absence of tetracycline. If, however, the selectable marker confers resistance to ampicillin, Kanamycin, or any of the other amino glycosides, you must dilute the transformant mix, say 1 to 10, into broth and incubate for 90-120 minutes prior to plating on selective media. In these cases you get few, if any, transformants if you plate immediately. This result, of course, is the same as reported by Cohen's lab several years ago.

S. WOO: How soon do you think an EK3 system will be available?

R. CURTISS: I surmise that it will be at least a year from now. In this country NIH has decided that all testing for certification of EK3 host-vectors will be done by contractors recruited by NIH. Since these contractors are just commencing their studies, I suspect it's going to be some time before they get all of the desired data. There will be five contractors testing each system to obtain information on survival in and on animals including primates, in waste water and during sewage treatment and as a consequence of aerosolization in laboratory settings and also on the phenotypic and genotypic properties of phage and plasmid vectors and their hosts.

S. COHEN: I recall that at Asilomar, a number of people felt that an EK2 host would be available in 6 weeks or less,of the meeting". Quite a bit of time has passed since then and I'm glad that Roy added in his last comment that an EK3 host would not be available for at least a year because the experience in the past has been that the development and approval of safer cloning systems has taken longer than anticipated.

R. CURTISS: I should add that the availability of a safer host-vector system is one thing, the approvability is another, χ 1776 in conjunction with pSC101 was finished in January 1976 with our testing data being submitted in April and was just officially approved six weeks ago. There have been numerous delays as people try to evaluate what EK2 means and what testing data is necessary. These delays have been understandable even if unfortunate. One can only hope that testing for EK3 certification and the evaluation of such testing data will proceed more smoothly.

THE NITROGEN FIXATION (*nif*) OPERON OF *Klebsiella pneumoniae*: CLONING *nif* GENES AND THE ISOLATION OF *nif* CONTROL MUTANTS

GERARD RIEDEL, ROBERT MARGOLSKEE, FRANK CANNON*, ADELE PESKIN,
and FRED AUSUBEL
Biological Laboratories
Harvard University
Cambridge, Massachusetts 02138

*A.R.C. Unit of Nitrogen Fixation
University of Sussex
Falmer, Brighton, Sussex BN1 9QJ U.K.

Abstract: We have constructed, *in vitro*, small amplifiable plasmids carrying *his* and *nif* genes from *Klebsiella pneumoniae*. One particular plasmid, pCRA37, complements mutations in *K. pneumoniae hisD*, *nifB* and *nifF* genes. The *K. pneumoniae* DNA fragment in pCRA37 is 10.9 megadaltons and was isolated as an *Eco*R1-generated partial digestion product of the stringent R'*nif* RP41.

We have also isolated *K. pneumoniae* mutants in which derepression of nitrogenase is independent of glutamine synthetase-mediated activation. These mutants, which we have designated *nif*T, map in the *his*-*nif* region of the genome. Expression of nitrogenase in *nif*T mutants is sensitive to ammonia repression, which indicates that nitrogenase expression is controlled by some factor(s) in addition to glutamine synthetase. This second controlling factor may be a specific *nif* repressor.

INTRODUCTION

Bacteria and blue green algae are the only organisms known to convert atmospheric nitrogen into a combined form such as ammonia. In all nitrogen-fixing species studied, N_2 is reduced to ammonia by a biochemically complex enzyme "Nitrogenase." The three dimensional structure of nitrogenase appears to have been conserved in evolution. All nitrogenases are composed of two characteristic subunits which can be dissociated and reassociated *in vitro* to form interspecies hybrids, some of which are enzymatically active (1).

The vast majority of terrestial nitrogen fixation occurs in the soil by free-living prokaryotes and in plants by prokaryotic symbionts. Industrial reduction of nitrogen in fertilizer factories probably accounts for less than 10% of the total amount fixed (2,3). Nitrogen fixation, whether catalyzed enzymatically by nitrogenase, or commercially by the Haber-Bosch process, requires high inputs of energy. As many as 15 moles of ATP are consumed for each mole of N_2 reduced to ammonia in *Clostridium pasteuriannum*, a free living nitrogen-fixing bacterium (4), and more than 2 x 10^6 barrels of oil are consumed per day world-wide in the manufacture of ammonia fertilizer (5).

The nif Gene Cluster of Klebsiella pneumoniae: The enzymology, physiology, and genetics associated with nitrogen-fixation are probably best understood in the bacterium *Klebsiella pneumoniae*. *K. pneumoniae* is a free-living (as opposed to symbiotic) nitrogen-fixing facultative aerobe, and has been chosen for study because many of the genetic techniques used in *E. coli* can also be used in *K. pneumoniae*. For example, cotransductional analysis, using the *E. coli* generalized transducing phage P1, has shown that a cluster of nitrogen-fixing (*nif*) genes maps between the operator end of the histidine (*his*) operon and the shikamate utilization genes (6-9). Figure 1 shows a map of the *his-nif* region of the *K. pneumoniae* chromosome which was drawn

...*hisD* *nifB* *nifF* *nifD,H* ...
0 5 10 15 20 25 30

Fig. 1. The *his-nif* region of the *K. pneumoniae* chromosome. The numbers below the line representing the *K. pneumoniae* chromosome are distances in kilo base pairs from *hisD*. The distances were calculated from cotransduction frequencies according to the method of Wu (11) from data in references 9 and 10, and on the basis of unpublished data obtained by F. Ausubel. Phenotypes of *nif* mutations: *nifB* mutants make defective molybdenum cofactor; *nifF* mutants are defective in nitrogenase specific electron transport; *nifD* mutants make defective component I of nitrogenase; and *nifH* mutants make defective component II of nitrogenase (9).

on the basis of data given in references 9 and 10 and on the basis of unpublished data obtained by F. Ausubel. The functions of the products of the pictured nif genes are described in the figure legend. We have calculated the distances in base pairs between several nif genes and the hisD gene using a formula derived by Wu (11) which converts cotransductional frequencies to physical distance on the genome. It is likely that all of the structural genes as well as any specific regulatory genes required exclusively for nif expression are located in this cluster, since transfer of this region to E. coli or Salmonella typhimurium by conjugation confers nitrogen-fixing ability upon these recipients (12-14).

The transfer of the K. pneumoniae nif gene cluster to E. coli (where it can be readily manipulated) has facilitated the construction of plasmids carrying nif genes. Cannon et al. (13) isolated an F' factor (FN68) in E. coli carrying the K. pneumoniae nif gene cluster, the his operon, and shiA. Dixon et al. (15) were then able to construct a promiscuous his nif shiA plasmid (RP41) derived by recombination between FN68 and the P-type drug resistance factor RP4. RP41 is a particularly useful genetic tool since it can be readily transferred to many, if not all, gram-negative bacteria (16). However, the large size of RP41 (at least 60 million daltons) and its low copy number per cell make it unsuitable for experiments which require the large scale isolation of purified nif DNA.

Ammonia Repression of Nitrogen Fixing Enzymes in K. pneumoniae: Because nitrogen-fixation is such an energy requiring process, it is not surprising to find that the synthesis of nitrogen-fixing enzymes in all organisms studied to date is repressed by the presence of a source of fixed nitrogen in the medium (17). We have been studying the mechanism by which ammonia coordinately represses the synthesis of enzymes of the nif gene cluster in K. pneumoniae. K. pneumoniae's preferred nitrogen source is ammonia, and several laboratories have been investigating the possibility that ammonia repression of several different enzymes is mediated by a common ammonia-sensing control center. In addition to nitrogenase, these enzymes include glutamine synthetase, which catalyzes the energy-requiring formation of glutamine from ammonia and glutamate, and the enzymes of the histidine utilization (hut) operon. Recently Magasanik and his coworkers (18) have identified glutamine synthetase as the common regulatory element mediating ammonia repression in K. aerogenes, a close relative of K. pneumoniae. They have postulated that under low ammonia conditions, glutamine synthetase acts as a positive transcriptional activator by binding specifically to promoters of operons which code for enzymes capable of providing the cell with glutamine or ammonia.

The regulatory function of glutamine synthetase was deduced from the observations that: 1) Both the form and the amount of glutamine synthetase change in response to changes in the ammonia concentration in the medium (19). 2) Mutants of *K. aerogenes* and *K. pneumoniae* which do not synthesize active glutamine synthetase (*glnA*⁻) do not derepress the *hut* operon during ammonia-limited growth in the presence of histidine, while mutants which produce glutamine synthetase constitutively (*glnC*⁻) are capable of synthesizing derepressed levels of the *hut* enzymes even in the presence of high levels of ammonia (20). 3) Purified glutamine synthetase activates the transcription of *hut*-specific mRNA *in vitro* when purified λ *hut* transducing phage DNA is used as a template for transcription (21).

Glutamine synthetase regulation of "nitrogen supply" enzymes (e.g., *hut*) can be considered to be formally analogous to catabolite activator protein (CAP) regulation of enzymes which provide the cell with energy. In each case, a single regulatory protein controls the expression of several different operons all of which play a similar physiological role in cellular metabolism (22).

Nitrogen-Fixing Genes in K. pneumoniae are Regulated by Glutamine Synthetase: Discovery of the regulatory role of glutamine synthetase in *hut* expression in *K. aerogenes* led to the postulate that glutamine synthetase may play a similar role in the regulation of nitrogenase expression in *K. pneumoniae*, since both enzymatic pathways serve the same purpose of providing the cell with nitrogen. This prediction appears to be accurate since preliminary experiments indicate that glutamine synthetase does, in fact, act as a positive controlling element in nitrogenase expression (23, 24): *glnA*⁻ mutants do not derepress nitrogenase during ammonia-limited anaerobic growth, and *glnC* mutants synthesize nitrogenase constitutively in the presence of ammonia (23).

* * * * *

Three experimental observations have guided the development of our particular strategy for elucidating the molecular mechanisms underlying *nif* control: 1) Glutamine synthetase mediates the regulation of nitrogenase, 2) The *nif* genes appear to cluster between *his* and *shiA*, and 3) The genes in the *nif* cluster are coordinately regulated. These observations led us to two conclusions: 1) It is likely that a single chromosomal site responds to glutamine synthetase-mediated activation and that this site could be identified by mutation. 2) It should be possible to clone all of the nitrogen-fixing functions in vitro on a single amplifiable plasmid. An amplifiable nif plasmid would allow us to purify large quantities of nif DNA which could be used in vitro in biochemical studies of nif regulation.

In this brief review of work on K. pneumoniae from our laboratory, we describe the in vitro construction of a plasmid containing some (but not all) of the genes from the nif gene cluster and the his operon. We also describe the isolation of mutants in which derepression of nitrogenase is independent of glutamine synthetase activation. These mutations map in the nif gene cluster and have phenotypes expected of promoter mutations.

Cloning Klebsiella pneumoniae his and nif Genes

Experimental Strategy: Figure 2 summarizes the strategy we employed to clone the nif gene cluster from K. pneumoniae (25). As a source of nif genes, we decided to use purified RP41 DNA. (RP41 is the R'nif described above which appears to carry all of the his-linked nif genes.) RP41 has a molecular weight of approximately 60 megadaltons (15) and thus contains nif genes in a highly concentrated form compared to K. pneumoniae chromosomal DNA. Although RP41 DNA is difficult to purify in large quantities, we were able to purify sufficient amounts for endonuclease restriction analysis and for in vitro molecular cloning experiments.

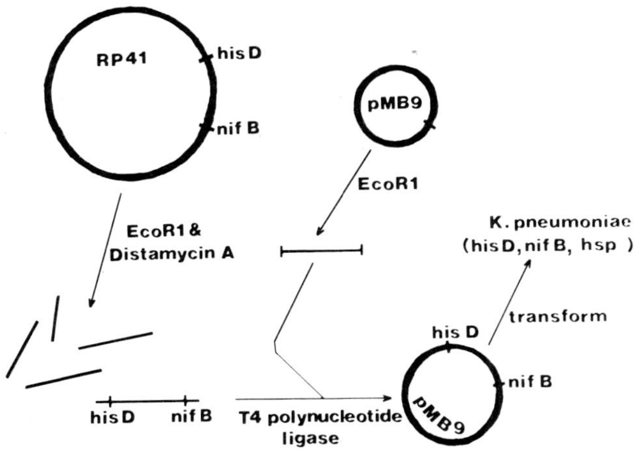

Fig. 2. Diagram of cloning strategy.

The experimental design shown in Figure 2 circumvents three problems we expected to encounter in the cloning of nif genes. The first problem is that a DNA fragment large enough to contain all or most of the genes in the nif cluster would probably contain more than one restriction site for many different restriction endonucleases. As shown in Figure 1, we estimated that a DNA fragment containing hisD and all of the his-linked nif genes would have to have a molecular weight of approximately 20 megadaltons (30 kilobases). In fact, the restriction endonucleases Sal1, EcoR1, HindIII, and Bam1 all cleave RP41 DNA into at least 10 fragments, most of which are considerably smaller than 20 megadaltons (25).

To decrease the possibility of cloning only small fragments of the nif cluster, we decided to clone partial restriction digests of RP41 DNA. This was accomplished by digesting RP41 DNA with EcoR1 in the presence of distamycin A, an antibiotic which binds to A-T rich regions of DNA and protects these regions from EcoR1 endonucleolytic activity (26,27). Figure 3 shows that EcoR1 cleavage of RP41 DNA in

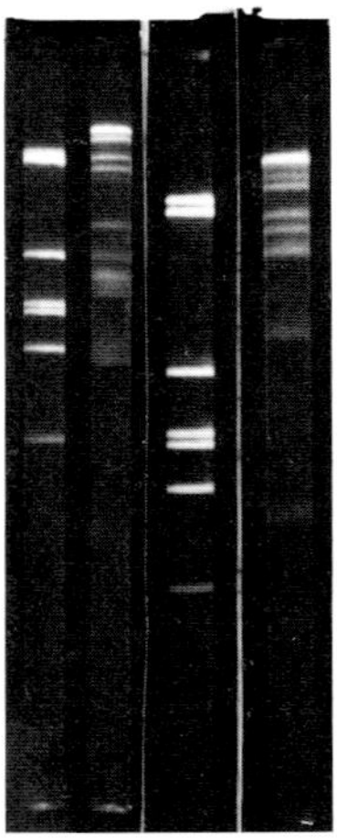

Fig. 3. Agarose gel electrophoresis of restricted RP41 DNA. A,C: Phage lambda DNA digested with EcoR1. The bands correspond to DNA fragments of molecular weights: 15.8, 13.7, 4.7, 3.7, 3.5, 3.0, 2.1 megadaltons. B: RP41 DNA digested with EcoR1. D: RP41 DNA digested with EcoR1 in the presence of 12 ug/ml distamycin A. Electrophoresis was for 14 hours at 1 mA/gel tube on 0.8% agarose (A,B) or on 0.65% agarose (C,D).

the presence of 12 ug/ml distamycin A generates 6 prominent fragments all greater than 6 megadaltons which are discernable over a background smear of DNA. The background appears to be a mixture of a number of different partial digestion products.

The second problem associated with cloning nif genes results from the inability to select efficiently for a Nif^+ phenotype on agar plates using N_2 as a sole nitrogen source. We have observed, for example, that direct selection for phage P1-mediated generalized transduction of nif alleles occurs at a considerably lower frequency than transduction of the $hisD^+$ allele in K. pneumoniae. We therefore decided to take advantage of the close linkage of hisD to the nif cluster and to select first for plasmids carrying the hisD allele and then subsequently screen these plasmids for ones which also carry nif genes.

The third difficulty in cloning the nif cluster is a consequence of the poor transforming ability of large plasmids. We have observed that plasmids greater than 16 megadaltons transform at an extremely low frequency and we expect a his-nif plasmid to be considerably larger than this. Thus, although one of our primary concerns was the inclusion of a procedure to generate DNA fragments large enough to contain the entire nif cluster, we also included a procedure to identify plasmids which only contain a portion of the his-linked nif genes. This was accomplished by using a recipient strain which carries a nif mutation (nifB) closely linked to hisD.

Plasmid Construction: We chose pMB9 as a cloning vehicle. This is a 3.5 megadalton amplifiable plasmid carrying tetracycline resistance genes, a ColE1 replicator, and a single recognition site for EcoR1 endonuclease (28). We ligated RP41 fragments generated by EcoR1 digestion in the presence of 12.5 ug/ml distamycin A to EcoR1 linearized pMB9; conditions were chosen which initially maximized end-to-end joining and subsequently favored circularization of DNA molecules. The ligated DNA was used to transform a hisD nifB hsp str derivative of K. pneumoniae strain M5a1. This recipient strain, KP5058, was chosen (as described in Figure 2) because: 1) The hisD mutation allowed direct selection for a His^+ phenotype instead of a Nif^+ phenotype. 2) The nifB mutation (80% cotransducible with hisD) allowed us to select for plasmids carrying only a portion of the nif cluster. 3) The hsp (restrictionless phenotype) prevented degradation of pMB9 and RP41 DNA which had been isolated from E. coli. We screened 140 His^+Tc^R transformants of KP5058 for nitrogenase activity using the acetylene reduction assay and found 9 that had a Nif^+ phenotype. All nine transformants contained a plasmid of 14.4 ± 0.3 megadaltons, and we chose one of the plasmids (pCRA37) for detailed study.

Characterization of pCRA37: Figure 4 shows the locations of a variety of restriction endonuclease recognition sites on pCRA37. The plasmid is composed of pMB9 and four EcoR1 fragments. We believe that the order of the EcoR1 fragments in pCRA37 is the same as the order of the same fragments on the chromosome, since two plasmids identical to pCRA37 were constructed in independent cloning experiments. The EcoR1 recognition sites between fragments D and E, between E and A, and between A and C were apparently protected from cleavage by distamycin A during EcoR1-digestion of RP41 DNA in the cloning protocol described above.

In Figure 4, the length of pCRA37 as determined by agarose gel electrophoresis is compared to the estimated length of the his-nif region of the K. pneumoniae genome, determined as described in Figure 1. The EcoR1 fragment, labeled

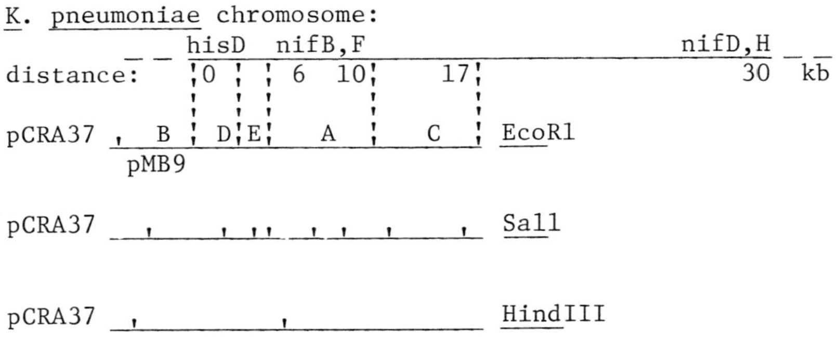

Fig. 4. Structure of pCRA37. Restriction map of pCRA37 drawn to scale with respect to the estimated length of the his-nif region of the K. pneumoniae chromosome as determined in Figure 1.

"D" (which is 1.8 megadaltons) contains the hisD gene, since a plasmid composed only of pMB9 and fragment D complements hisD mutations in E. coli and K. pneumoniae. This places the right hand end point of pCRA37 to the right of nifF and to the left of the nitrogenase structural genes, assuming that the estimated length of the nif gene cluster shown in Figures 1 and 4 is accurate. This leads to the prediction that pCRA37 should complement nifB and nifF mutations but should not complement mutations in the nitrogenase structural genes.

The data in Table 1 demonstrate that pCRA37 complements a nifB mutation, a nifF mutation, and nif deletions which extend from his through nifF. On the other hand, pCRA37

TABLE 1

Transformation of K. pneumoniae nif^- strains with pCRA37 plasmid DNA

Exp	Recipient	Recipient genotype	Transformant phenotype
1	KP5058	hisD nifB	35/39 Nif^+,39/39 Tc^R
2	KP5058	hisD nifB	20/20 Nif^+,20/20 Tc^R
3	UN150	hisD nifB	14/14 Nif^+,14/14 Tc^R
4	UN587	hisD nifF	7/7 Nif^+, 7/7 Tc^R
5	UN906	his-nifB-nifF (deletion)	16/16 Nif^+,16/16 Tc^R
6	UN901	his-nifB-nifD (deletion)	0/16 Nif^+,16/16 Tc^R
7	C603 (E. coli)	hisD (nif^-)	0/120 Nif^+,120/120 Tc^R

does not complement deletions extending from his through the gene (nifD) which codes for component I of the nitrogenase enzyme. The conclusion that pCRA37 does not carry the nitrogenase structural genes is corroborated by the observation that pCRA37 transformants of E. coli do not have nitrogenase activity whereas RP41 transconjugants of E. coli have the same level of nitrogenase activity as wild type K. pneumoniae.

Examination of Table 1 reveals an interesting phenomenon: Only 90% of the His^+ Tc^R transformants of KP5058 obtained with pCRA37 in a particular experiment were also Nif^+. To determine why a low but significant proportion of the transformants were Nif^-, we isolated plasmid DNA from one of the His^+ $Tc^R Nif^-$ transformants. DNA isolated from this strain yielded two plasmid species: pCRA37 (approximately 5% of the total plasmid weight) and pCRA111 (approximately 95% of total plasmid weight). A preliminary restriction characterization of pCRA111 showed that a large part of pCRA37 had been deleted, generating a 4.9 megadalton plasmid which contained part of pMB9, a part of EcoR1 fragment E and all of EcoR1 fragment D. Plasmid breakdown of this kind could explain the appearance of

His^+ Tc^RNif^- transformants and we are studying the phenomenon in greater detail.

One observation that we have made suggests an alternative explanation for Nif^- pCRA37 transformants. When KP5058/pCRA37 is cultured in the presence of 10 ug/ml tetracycline and then individual clones are examined for nitrogenase activity, only 1-2% of the clones are Nif^+. We have tentatively concluded that derepression of the tetracycline resistance genes on pCRA37 interferes with *nif* transcription. In contrast, KP5058/RP41 displays a Nif^+ phenotype irrespective of whether the strain is grown in the presence of tetracycline.

We have also examined the spontaneous loss of pCRA37 from KP5058. After one subculture in nutrient broth, 5 out of 30 clones tested were $His^-Tc^SNif^-$, and no plasmid DNA was detectable in these clones. The remaining 25 clones were $His^+Tc^RNif^+$ and all contained pCRA37. These results indicate that the *nif* and *his* genes on pCRA37 are complementing corresponding mutations in KP5058 and decreases the likelihood that the Nif^+His^+ phenotypes of pCRA37 transformants result from recombination between pCRA37 and the chromosome. We are currently attempting to construct recombination-deficient *K. pneumoniae* mutants for future complementation studies.

Discussion of pCRA37: It is possible that the length of the *nif* gene cluster determined from cotransductional frequencies is overestimated since the accuracy of the calculation in *K. pneumoniae* depends on the assumption that the recombination frequency is the same in *E. coli* and in *K. pneumoniae*. Thus, pCRA37 may actually contain the nitrogenase structural genes and the failure of pCRA37 to complement nitrogenase structural gene mutations may be due to the lack of an active promoter on pCRA37 for these genes. The fact that pCRA37 complements *nif*B and *nif*F mutations implies that the *nif* gene cluster (as depicted in Figure 4) is transcribed from left to right, that the *nif* gene cluster contains internal promoters, or that the *nif*B and *nif*F genes are transcribed from a promoter on pMB9. We are currently attempting to determine whether expression of the *nif*B genes on pCRA37 is under glutamine synthetase regulatory control as would be expected if the *nif*B gene were being transcribed from a *nif* specific promoter.

Assuming that pCRA37 does not contain the nitrogenase structural genes, our primary goal will be to clone the remaining *his*-linked *nif* genes. If our size estimate of the *nif* cluster is correct, it will be difficult to clone the entire *nif* gene cluster in a single *in vitro* experiment. To circumvent this difficulty, we will clone the *nif* cluster in two pieces on nonhomologous cloning vehicles. If these two plasmids share a region of homology near the middle of the nif cluster, a complete nif gene cluster on a single plasmid can be constructed in vivo by recombination as described by Bedbrook and Ausubel (29).

REGULATORY MUTANTS IN THE nif GENE CLUSTER

We have argued above that a single chromosomal site in the nif gene cluster responds to glutamine synthetase activation. A prediction of this model is that it should be possible to isolate mutations in a nif promoter which allow transcription of the nif genes independent of glutamine synthetase activation. These would be analogous to the E. coli lactose operon mutation UV5 (30) in which lac transcription is independent of catabolite activator protein activation.

Experimental Plan: Strains of K. pneumoniae carrying glnA$^-$ mutations are phenotypically Nif$^-$ and do not derepress nitrogenase when starved for ammonia. There are two reasons, however, why glnA mutants cannot be used directly to select Nif$^+$ revertants containing mutations in a nif promoter. Ammonia formed by nitrogenase-catalyzed reduction of N_2 is assimilated into intermediary metabolic pathways via glutamine synthetase-catalyzed formation of glutamine. Thus, a glnA$^-$ strain also carrying a nif mutation which allowed transcription of the nitrogenase genes in the absence of glutamine synthetase activation might be unable to assimilate ammonia formed by nitrogenase. Also, glnA mutants must be supplemented with relatively high levels of glutamine in order to satisfy the glutamine auxotrophy. This level of glutamine provides the cell with sufficient nitrogen to grow without the necessity of reducing N_2 to ammonia.

The strategy we developed to circumvent these difficulties in the selection of Nif$^+$ revertants in glnA$^-$ strains depends on the finding that glutamine synthetase appears to have at least two important roles in the cell. First, it is an essential catalyst in the biosynthesis of glutamine, and second, it serves in a regulatory capacity, stimulating nif and hut transcription. There was no a priori reason why these two activities could not be mutated independently to generate a strain which lacked the ability to activate hut and nif transcription but retained the capacity to synthesize glutamine. We have adopted the nomenclature "glnA" and "glnR" to designate the biosynthetic and regulatory roles of glutamine synthetase respectively. A mutant with only biosynthetic activity would be designated glnA$^+$glnR$^-$ (glnA$^+$R$^-$) and it would have the phenotype $Gln^{+}Nif^{-}Hut^{-}$. If this mutant strain could be isolated, it could be used directly to select Nif^{+} revertants for two reasons: 1) It would not have to be supplemented with glutamine. 2) The ammonia formed from N_2 reduction could be assimilated and used as a sole nitrogen source.

<u>Isolation of $glnA^{+}glnA^{-}$ Strains</u>: The following strategy was used to obtain mutants with the $glnA^{+}R^{-}$ genotype (31): Revertants of a $glnA^{-}R^{-}$ strain (a glutamine auxotroph with the phenotype $Gln^{-}Nif^{-}Hut^{-}$) were selected for a Gln^{+} phenotype. These Gln^{+} revertants fell into two classes as shown on lines 2a and 2b of Figure 5. The first class

step	genotype	phenotype			linkage of Nif+ phenotype
		Gln	Nif	Hut	
1	$glnA^{-}R^{-}_{nif}R^{-}_{hut}$	−	−	−	met
2a	$glnA^{+}R^{+}_{nif}R^{+}_{hut}$	+	+	+	−
2b	$glnA^{+}R^{-}_{nif}R^{-}_{hut}$	+	−	−	met
3a	$glnA^{+}R^{+}_{nif}R^{+}_{hut}$	+	+	−	−
3b	$glnA^{+}R^{+}_{nif}R^{-}_{hut}$	+	+	−	met
3c	$glnA^{+}R^{-}_{nif}R^{-}_{hut}$ $nifT^{-}$	+	+	−	his

Fig. 5. Strategy used to isolate nif mutants which are independent of glutamine synthetase activation. Strains shown on lines 2a and 2b were obtained from strain 1 by selecting Gln^{+}. Strains on lines 3a, 3b, and 3c were obtained from strain 2b by selecting Nif^{+}.

appeared to have regained wild-type glutamine synthetase biosynthetic and regulatory activities. The second class regained the ability to synthesize glutamine, but appeared unable to activate nif and hut transcription. This phenotype is what we expected of a $glnA^{+}R^{-}$ mutant and we chose one such mutant, KP502-3, for further study.

In order to demonstrate that the Nif^{-} and Hut^{-} phenotype of strain KP502-3 was due to (a) mutation(s) in the *glnA* gene, we mapped the GlnR$^-$ phenotype. We have shown that in *K. pneumoniae* (31), the *glnA* gene is 4% linked to *metB* and 7-12% linked to *rha*. We found that the GlnR$^-$ phenotype in strain KP502-3 has the same linkage to these markers as the *glnA* gene. Furthermore, the GlnR$^-$ phenotype is 100% linked to *glnA*, and is not linked to the *his-nif* region of the chromosome. These data demonstrate that the GlnR$^-$ phenotype is most likely due to (a) mutation(s) in the glutamine synthetase structural gene.

Table 2 shows that strain KP502-3 produces only 2% of wild-type nitrogenase activity at non-repressing levels of ammonia. This level of activity was sufficiently low that KP502-3 could be used to select Nif$^+$ revertants directly on ammonia limited agar medium. Although KP502-3 is phenotypically Nif$^-$ under our normal nitrogenase assay conditions and on agar plates, we have recently discovered that nitrogenase in KP502-3 can be derepressed to wild-type levels if special efforts are made to reduce the ammonia concentrations to very low levels in the growth medium. The significance of this result is discussed at length below.

Isolation of nif Mutants which are Independent of Glutamine Synthetase Activation: Two different mutagenesis procedures produced phenotypic Nif$^+$ revertants of strain KP502-3 (glnA$^+$R$^-$) (31). In the first procedure, Nif$^+$ revertants were selected from a nitrosoguanidine (NTG) mutagenized culture. A total of 48 independent Nif$^+$ revertants were screened for Hut$^\pm$ phenotype. Of the 48 Nif$^+$ revertants 43 were Hut$^+$. We assumed that these Nif$^+$Hut$^+$ revertants were due to mutations in glutamine synthetase which restored wild-type regulatory activity (line 3a, figure 5). The five remaining Nif$^+$Hut$^-$ revertants were tested further as presumptive *nif* mutants, independent of glutamine synthetase. However, phage P1 mediated cotransductional mapping localized the mutation responsible for the Nif$^+$ phenotype to the *glnA* region of the chromosome. The data in Table 2 show that in two such mutants (KP507-2 and KP507-4) the Nif$^+$Hut$^-$ phenotype is 100% cotransducible with *glnA* and 8-12% cotransducible with *rha*. The other three mutants gave similar results. These experiments indicate that the glutamine synthetase regulatory activities for *nif* and *hut* could be mutated independently. According to the nomenclature we adopted above, these mutants would have the genotype $glnA^+R^+_{nif}R^-_{hut}$ as indicated on line 3b in Figure 5.

It is not surprising that Nif$^+$ revertants which mapped only in the *glnA* region were obtained when Nif$^+$ revertants of KP502-3 were selected directly. The *glnA* gene is probably 2-3 orders of magnitude larger than the *nif* promoter region containing the postulated glutamine synthetase binding site. Rather than screening hundreds of Nif^+ revertants of KP502-3 in the hope of finding nif promoter mutants, we decided to enrich specifically for mutations linked to the his-nif region of the chromosome. We used a procedure developed by Oeschger and Berlyn (32) which takes advantage of the fact that NTG causes closely linked multiple mutations at the replicating forks of the chromosome. In *K. pneumoniae*, *his* and *nif* are closely linked whereas *gln* is unlinked to either *his* or *nif*. Therefore, since KP502-3 is $hisD^-$, by selecting simultaneous His^+Nif^+ revertants of NTG mutagenized KP502-3, we enriched for Nif^+ revertants linked to *hisD*. Two independent $Nif^+His^+Hut^-$ revertants obtained in this manner (KP5160-3 and KP5161-3) were chosen for further study. The data in Table 2 show that the Nif^+ phenotype in these strains is due to a mutation which is cotransducible with *hisD*. We have designated this new mutation "*nifT*". We have also demonstrated that these two *nifT* mutants contain the original $glnA^+R^-$ mutation. This Nif^+ revertant class of KP502-3 is represented on line 3c in Figure 5.

The $nifT^-$ mutation in strain KP5161-3 is dominant as expected for a promoter mutation. We are currently performing tests to determine whether the *nifT* mutations are *cis* or *trans* dominant. One unexpected property of the $nifT^-$ mutants is also shown in Table 2. Nitrogenase in the two $nifT^-$ strains is completely repressed by 0.2% ammonia. The significance of this result is described in detail below.

Discussion of Mutations Affecting nif Regulation: Two important conclusions can be drawn from the data presented here: 1) The hypothesis that glutamine synthetase plays a direct role in the transcriptional regulation of the *hut* and *nif* operons (18,22) is supported by the isolation of glutamine synthetase mutations which exhibit differential regulatory phenotypes with respect to *nif* and *hut* control ($glnA^+R^+_{nif}R^-_{hut}$). 2) The *nif* operon in *K. pneumoniae* is probably subject to control by a second ammonia sensing element independent of glutamine synthetase. This second conclusion derives from the observation that nitrogenase in $nifT^-$ mutants is repressible by ammonia and contradicts the hypothesis that glutamine synthetase is the sole regulatory element for *nif* expression (22,23).

One possible mechanism for the second type of control is a *nif* specific repressor which, in the presence of ammonia, binds to a *nif* operator and blocks *nif* transcription. If this were the case, the *nif* gene cluster would be subject to both positive and negative controls. According to this model, at high ammonia concentrations, the transcriptional activator (glutamine synthetase) would be inactive, and the repressor would be fully active. The result is complete repression of nif transcription. At low ammonia concentrations, glutamine synthetase would activate transcription and the repressor would be inactivated. Within some specific range of intermediate ammonia concentrations, the activator and repressor would be competing for the control of nif transcription.

This model of two competing controlling elements for nif transcription can be used to explain the phenotypes of all of the mutations affecting nif expression which have been isolated in our laboratory. Strains with a glnA$^+$R$^-$ genotype lack nif activator and thus nif expression is controlled solely by the repressor. As we have mentioned above, such strains can be derepressed for nitrogenase at very low ammonia levels. This result indicates that glutamine synthetase activation is not an absolute requirement for nif transcription, but instead acts to modulate the repressor's control. Thus a glnA$^+$R$^-$ strain would be expected to be much more sensitive to ammonia repression than a wild-type strain but could be derepressed by inactivating the repressor at extremely low levels of ammonia.

Previously, the observation that glutamine auxotrophs (glnA$^-$R$^-$) were Nif$^-$ was interpreted by postulating an absolute requirement for glutamine synthetase activation for nif expression (23). In the light of new evidence presented here, however, a more plausible explanation for the Nif$^-$ phenotype is that the glnA$^-$ genotype results in the intracellular accumulation of ammonia since glnA$^-$ strains must be supplemented with glutamine and the resulting ammonia concentration is sufficient to keep the nif repressor fully activated.

The nifT mutation probably results in a low level of constitutive transcription at a nif promotor which is insensitive to the nif repressor at intermediate levels of ammonia. At this same ammonia concentration, a glnA$^+$R$^-$ strain is completely repressed. At higher levels of ammonia, however, the repressor overcomes this constitutive level of transcription and nif is repressed.

The glnC mutation results in a high continuous level of activation, even in the presence of high ammonia levels, which overwhelms the repressor.

Our model of nif control leads to the prediction that mutations in the presumptive nif repressor would lead to nif constitutivity in the presence of ammonia. We are in the process of isolating such mutants.

REFERENCES

(1) R.W. Detroy, D.F. Witz, R.A. Parejko and P.W. Wilson, Proc. Natl. Acad. Sci. U.S.A., 61 (1968) 537.

(2) R.C. Burns and R.W.F. Hardy, Nitrogen Fixation in Bacteria and Higher Plants (Springer, New York, 1973).

(3) J.R. Postgate, J. Appl. Bact., 37 (1974) 185.

(4) H.C. Winter and R.H. Burns, J. Biol. Chem, 243 (1968) 940.

(5) G.C. Sweeney, in: Proceedings of the First International Symposium on Nitrogen Fixation, Vol. 1, eds. E.W. Newton and C.J. Nyman (Washington State University Press, Washington).

(6) S.L. Streicher, E.G. Gurney and R.C. Valentine, Proc. Natl. Acad. Sci., U.S.A., 68 (1971) 1174.

(7) R.A. Dixon and J.R. Postgate, Nature, 234 (1971) 47.

(8) K.L. Shanmugam, A.S. Loo and R.C. Valentine, Biochim. Biophys. Acta, 338 (1974) 545.

(9) R.T. St. John, H.M. Johnston, C. Seidman, D. Garfinkel, J.K. Gordon, V.K. Shah and W.J. Brill, J. Bacteriol., 121 (1975) 759.

(10) S.L. Streicher, E.G. Gurney and R.C. Valentine, Nature, 239 (1972) 495.

(11) T.T. Wu, Genetics, 54 (1966) 405.

(12) R.A. Dixon and J.R. Postgate, Nature, 237 (1972) 102.

(13) F.C. Cannon, R.A. Dixon and J.R. Postgate, J. Gen. Microbiol., 93 (1976) 111.

(14) F.C. Cannon and J.R. Postgate, Nature, 260 (1976) 271.

(15) R.A. Dixon, F.C. Cannon and A. Kondorosi, Nature, 260 (1976) 268.

(16) N. Datta, R.W. Hedges, E.J. Shaw, R.B. Sykes and M.H. Richmond, J. Bacteriol., 108 (1971) 1244.

(17) H. Dalton and L.E. Mortenson, Bact. Rev., 36 (1972) 231.

(18) B. Magasanik, M.J. Prival, J.E. Brenchley, B.M. Tyler, A.B. DeLeo, S.L. Streicher, R.A. Bender and C.G. Paris, in: Current Topics in Cellular Regulation, Vol. 8, eds. L. Horecker and E.R. Stadtman, (Academic Press, New York, 1974) p. 119.

(19) M.J. Prival and B. Magasanik, J. Biol. Chem., 246 (1971) 6288.

(20) M.J. Prival, J.E. Brenchley and B. Magasanik, J. Biol. Chem., 248 (1973) 4334.

(21) B.M. Tyler, A.B. DeLeo and B. Magasanik, Proc. Natl. Acad. Sci. U.S.A., 71 (1974) 225.

(22) B. Magasanik, in: Progress in Nucleic Acid Research and Molecular Biology, Vol. 17 (Academic Press, New York, 1976) p. 99.

(23) S.L. Streicher, K.T. Shanmugam, F.M. Ausubel, C. Morandi and R. Goldberg, J. Bacteriol., 120 (1974) 815.

(24) R.S. Tubb, Nature, 251 (1974) 481.

(25) F.C. Cannon, G.E. Riedel and F.M. Ausubel, submitted for publication.

(26) E.A. Braga, V.V. Nosikov, V.I. Tanjashhin, A.L. Zhuze and O.L. Poljanovskij, Dok. Akad. Nauk. USSR, 225 (1975) 707.

(27) V.V. Nosilov, E.A. Braga, A.V. Karlishev, A.L. Zhuze and O.L. Poljanovskij, Nuc. Acid Res., 3 (1976) 2293.

(28) R.L. Rodriguez, F. Bolivar, H.M. Goodman, H.W. Boyer and M.C. Betlach, in: ICN-UCLA Symposium on Genetic Regulatory Mechanisms, (Academic Press, New York) in press.

(29) J.R. Bedbrook and F.M. Ausubel, Cell, 4 (1976) 707.

(30) R.R. Arditti, J.G. Scaife and J.R. Beckwith, J. Mol. Biol., 38 (1968) 421.

(31) F.M. Ausubel, R.M. Margolskee and N. Maizels, in: Proceedings of the 2nd International Symposium on Nitrogen Fixation, (Academic Press, New York) in press.

(32) M. Oeschger and M. Berlyn, Molec. Gen. Genet., 134 (1974) 77.

ACKNOWLEDGEMENTS

We would like to thank Kaaren Janssen and Stephanie Bird for constructive criticism of the manuscript. The work on nif regulatory mutants was conducted at the Biological Laboratories, Harvard University; and the work on nif cloning was conducted primarily at the Biological Laboratories, Harvard University and in part at the A.R.C. Unit of Nitrogen Fixation, University of Sussex. This work was supported in part by a National Science Foundation Grant No. PCM75-21435 A01 to F.A.

DISCUSSION

R.H. LAWRENCE: Would you elaborate on current experiments in your or in other laboratories that are leading towards the transfer and expression of this uniquely prokaryotic genetic information in eukaryotic cells (especially plant cells)?

F. AUSUBEL: As far as I am aware there are no authenticated examples of both transcriptions and translation of any prokaryotic gene in eukaryotic cells. Experiments that have been published which appear to indicate this are not particularly convincing.

R.H. LAWRENCE: Then in view of the data you presented today what is your research scheme to approach such a genetic transfer?

F. AUSUBEL: In the first place, in order for prokaryotic genes to be expressed in a eukaryotic host they have to be replicated in some efficient way, and that probably is going to involve associating them with a eukaryotic replicon, for example, a plant or animal virus. Secondly, it is unlikely that the prokaryotic transcriptional and translational signals will be recognized in eukaryotic cells. Special genomes will probably have to be engineered on which eukaryotic signals are placed adjacent to the prokaryotic structural genes. But I think it is premature to think about doing experiments like this at the present time. We have to wait until a little more is known about the structure of eukaryotic genes.

APPLICATIONS OF BACTERIOPHAGE λ IN RECOMBINANT DNA RESEARCH

KENNETH MURRAY
Department of Molecular Biology,
University of Edinburgh,
Edinburgh EH9 3JR
Scotland, U.K.

The attractions of the bacteriophage λ genome as a receptor molecule for the construction of DNA recombinants *in vitro* stem from the extensive background knowledge and experience of the genetics of the virus (1,2). This can be exploited not only in the construction of a range of receptors for specific needs, but also in providing procedures for the efficient recovery of recombinant DNA molecules from the biochemical reactions employed in their production, in the development of systems that facilitate the discrimination between recombinant and parental phages or permit selection of the recombinants, in manipulations leading to the optimal expression of inserted DNA sequences, and in the detailed analysis of the inserted sequences themselves.

1. The Development of Receptor Chromosomes for DNA Fragments

The principal desiderata for a DNA molecule that is to serve as a receptor, or vector, for DNA fragments formed by the action of restriction enzymes are the following. It should be small, easily prepared and able to replicate autonomously in an appropriate bacterial host. It should contain only one target for the restriction enzyme used to generate the DNA fragments of interest (or a small number of targets if these enclose or are within non-essential or replaceable regions of the receptor DNA). It should carry some simple characteristic (such as drug resistance, immunity, plaque formation, or a gene permitting its recognition by complementation of an auxotrophic strain) by which transformed cells may be readily recognised. Insertion of a DNA fragment into the receptor obviously must not destroy an essential function. Finally some simple distinction between the receptor and the recombinant phenotypes is desirable.

The target sequences for the majority of the restriction endonucleases used in the *in vitro* construction of recombinant DNA molecules are hexanucleotides. Any given hexanucleotide sequence will occur by chance once in 4^6, or 4096, bases in a polynucleotide containing equal proportions of the four standard bases. Bacteriophage lambda particles contain one linear, double-stranded DNA molecule with a molecular weight of about 32 x 10^6, which is equivalent to about 45,000 base pairs, so that a target consisting of a hexanucleotide sequence would occur by chance about eleven times in the molecule. However, the number of target sequences that actually occur in the λ genome is not, of course, the same as the number expected by chance, and in many cases is appreciably fewer. The wild-type lambda genome has five targets for the R.*Eco*RI enzyme[1] (6), six targets for the R.*Hind*III enzyme (7), but only one for R.*Xho* and R.*Xba* (4).

Since the phage DNA molecule must be enclosed within a capsid of defined size (8) there are distinct limitations upon the size of this molecule and in practice stable λ virions are only formed with DNA molecules ranging from about 75% to 106% of the size of wild-type λ (i.e. λ^+) DNA (9). A large number of naturally occurring deletion mutants of λ have been obtained and the positions of the deletions within the DNA molecule have been determined by heteroduplex electron microscopy (1).

Deletion mutants provide the starting point for the construction of lambda genomes that are to serve as receptor chromosomes for the construction and propagation of recombinant DNA molecules. Not only do they provide space for the DNA fragment that is to be incorporated into the phage genome, but they also offer one method for removing restriction targets from the genome, either from positions that may be inconvenient, or simply to reduce the number of DNA fragments to be rejoined in the construction of the hybrid DNA molecule.

Obviously, the location within the lambda genome of the target sequences for a restriction enzyme must be determined in order to make a suitable receptor molecule for use with DNA fragments from that particular enzyme. There are

---

Footnote:

[1]The abbreviated description of restriction endonucleases is that suggested by Smith and Nathans (3). A more extensive list of restriction enzymes and their target sequences will be found in references 4 and 5.

now a number of ways of doing this. The classical approach to sequence determination based upon the use of partial digests which was used so successfully to map restriction targets in Simian virus 40 (SV40) DNA (10) is limited to the smaller viral genomes. In all cases, the problem is greatly simplified by the use of gel electrophoresis (acrylamide or agarose) both to separate fragments in restriction enzyme digests and simultaneously to estimate the sizes of the fragments; this may be done from measurements of electrophoretic mobility in gels calibrated with suitable standards and from measurement of relative yields of radioactivity in fragments from isotopically labelled DNA preparations (11). Fragment sizes may also be obtained from contour length measurements on electronmicrographs (6,12). These methods were used in conjunction with a number of known deletion and substitution derivatives of phage λ to locate the targets for the EcoRI (6) and HindIII enzymes (7). Two additional methods have been introduced and are particularly useful for mapping some of the smaller fragments (or targets that occur relatively frequently). One of them is based upon the extension of a selected fragment by reactions with DNA polymerase and nucleoside triphosphates after annealing the fragment (i.e. one of its strands) to a single strand of the parent DNA molecule (13). The other makes use of partial digests of fragments that have been terminally labelled; the reaction for labelling 5' termini by means of polynucleotide kinase is particularly convenient (14).

In addition to the use of deletion mutants, the distribution of restriction targets in phage genomes may be changed through substitution of appropriate regions of a related phage genome by means of a suitable genetic cross, or in some cases by selection of mutations leading to the loss of a target. The development of receptors for use with DNA fragments generated by the action of the EcoRI and HindIII enzymes illustrate the use of these procedures.

Figure 1 shows the location of the targets in λ^+ DNA for these two restriction enzymes (6,12,17), and the positions of a number of deletions and substitutions (1). The λb538 deletion clearly removes targets 1 and 2 for R.EcoRI and targets 1, 2 and 3 for HindIII, while a deletion made by removing the DNA between srIλ-1 and srIλ-2 in reactions *in vitro* would remove targets 1 and 2 for R.HindIII, but leave shnλ-3 intact.[1] When the number of targets in the DNA

Footnote: [1]The convention for description of restriction targets, or sites, is essentially as suggested by Arber (15) and is used and described more fully in ref. 16.

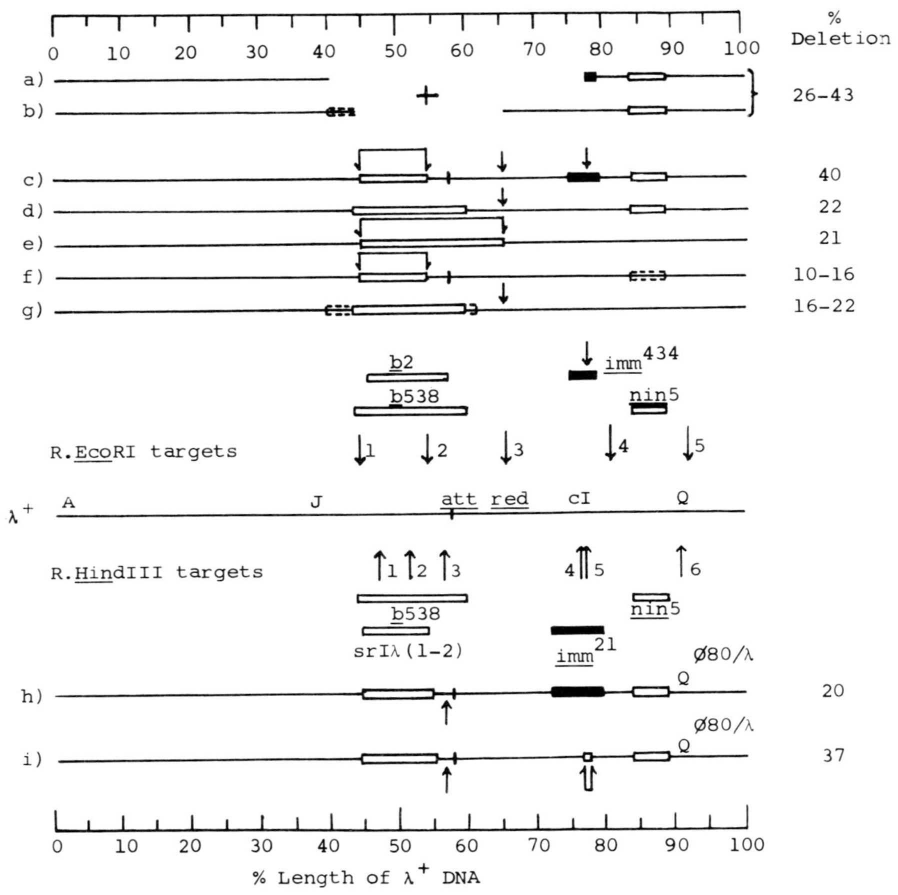

Fig. 1. Adaptation of the lambda genome to form receptors for DNA fragments. The vertical arrows above and below the line indicating the lambda DNA molecule show the positions of targets for the restriction enzymes EcoRI and HindIII, respectively. Open rectangles show the positions of deletions, the broken lines indicating the extent of alternative deletions that can be used to generate more space in the chromosome. Filled rectangles indicate the substitution of lambda by DNA from phages 434 and 21 (in each case the substitution is accompanied by a deletion). The short, vertical bar indicates the site, att, where the phage is integrated into the host chromosome upon lysogenisation. Phages with genomes d, e, f, g or h are simple insertion receptors, but those with combinations of fragments from a and b, or with c or i are replacement receptors and can accommodate larger fragments.

molecule becomes small manipulation of the remaining sites can be attempted through simple mutation, either by treatment with a mutagen or by selection of natural mutants.

If the restriction system in use is determined by cells which are a host for lambda, as with the EcoRI system, the selection of mutations removing a target is straightforward. The efficiency of restriction of a phage is measured simply as the ratio of titres of the unmodified phage on restricting and non-restricting strains and is dependent upon the number of targets in the phage DNA (17,18). Thus a mutation leading to loss of one of the EcoRI targets in a λb538 strain would leave a phage with two targets for R.EcoRI which would be less efficiently restricted than its parent on transfer from a non-modifying (RI) strain to a strain carrying the RI plasmid and the mutant would therefore have a selective advantage and so become enriched in the population. After a few cycles between the two host strains the mutant will comprise the greater proportion of the population so that only a few individual plaques need be tested in order to isolate the mutant. The remaining restriction targets may then be removed successively by repetition of the cycling between strains. These procedures were used successfully to remove from λ DNA targets for the EcoK system (18) and to provide phage genomes that serve as receptors for fragments of DNA digested with R.EcoRI (2), examples of which are in Fig. 1 (d and g), and also phages that have lost all targets for the enzymes (2,18). The latter are valuable for manipulation of targets from other, related, genomes by genetic crosses. For example, crosses with a λ derivative carrying the immunity of phage 434 enabled phages to be made which have single targets for the EcoRI or HindIII enzymes in the phage's cI gene and these are particularly useful receptors (16). Many λ derivatives have now been adapted for use as receptors for DNA fragments made by digestion with R.EcoRI (2,16,19,20).

For restriction systems carried by bacteria that are not hosts for the phage, the selection of mutated restriction targets, natural or induced, is still possible, but obviously more complex. Here DNA is prepared from the phage population, digested with the restriction enzyme in question and then used to transfect competent host cells. Phage with resistant genomes accumulate in the population and a large number of cycles of these operations may be necessary before phage with the desired genotype are recovered.

In adapting the λ chromosome as a receptor for DNA fragments released on digestion with R.HindIII, targets 1 and 2 for this enzyme were removed by means of the deletion *in vitro* of the region between srIλ-1 and 2 and targets 4 and 5, which were mapped in immunity, could be removed through the use of a deletion in this region, or by substituting the immunity of phage 21 for that of λ. Targets 3 and 6 then remained and 3 is clearly the more useful as the point at which to insert additional DNA. Substitution of part of the phage phi80 genome for that of λ in the region of λ gene Q removed shnλ-6 but left a functional gene Q so that the resulting phage (into which the nin5 deletion was introduced to make more space for the new fragment) could be used as a receptor (7), (Fig. 1h).

These and other examples in Fig. 1 show how the λ genome may be manipulated to change the distribution of restriction targets, and a large variety of receptors for various purposes and different restriction enzymes now exist or can be produced relatively easily (2,7,16,19,20,21). The introduction of targets by means of substitutions through genetic crosses is no less useful than the removal of targets and this may also be effected *in vitro*. For example, a phage DNA that is devoid of targets for a given restriction enzyme may contain two suitably placed targets for a second enzyme so that the DNA between these targets may be removed and a new DNA fragment from some other source which contains the desired target sequence may then be inserted in its place. The inserted fragment may be obtained from another phage or plasmid by digestion with the second restriction enzyme, or it may be a synthetic product. In a similar way, adaptor fragments may be used to alter the nucleotide sequence at any particular receptor site and hence change the specificity to that of another enzyme; this may be done with natural or synthetic fragments (22), but efficiencies of recovery of the recombinant DNA molecules are low with these complex products involving joinings of several fragments.

2. Recovery of Recombinant DNA as Phage.

Transfection of competent preparations of *E.coli* (prepared by starvation in 0.1 *M* $CaCl_2$; 23, 24) is the procedure most commonly employed for recovery and initial propagation of recombinant DNA molecules made in a sequence of restriction and ligation reactions, and this also provides a very convenient fractionation of the desired DNA molecules from the rather complex mixture of products in these reaction mixtures. The yields of phage are very variable and strain-dependent, but should normally be in the range of 10^5 to 10^6 plaque-forming units per microgram of DNA for intact molecules; with the reaction mixtures they are of course much less than this.

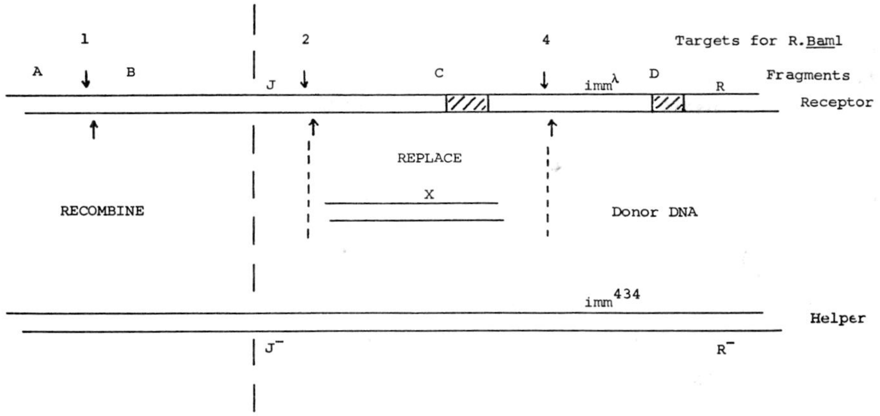

Fig. 2. The use of recombination with a helper phage to enhance the recovery of the in vitro recombinant DNA. The genome of the receptor phage (upper) retains targets 1, 2 and 4 for R.BamI. Correct joining of fragments A, B and D gives a molecule too small for packaging, so that fragment C or a replacement, X, of suitable size is essential for viability. Viable phage genomes of the type A, B, X, D, are formed by in vitro joining of the four fragments or by recombination left of the broken line between the partial genome B.X.D and the helper phage and are recognised as J^+ spi$^-$ immλ R^+ phages.

Helper mediated transformation of E.coli with λ is an alternative means for recovery of viable DNA molecules (25) and although somewhat more involved than simple transfection it is capable of giving appreciably higher yields of phage. A feature of this system is that the genome of the helper phage may be used to rescue incomplete recombinant molecules from the reaction mixtures by recombination in vivo. An example of this is illustrated in Fig. 2. The λ genome has five targets for R.BamI (26,27,28), four of which may be readily removed but sbmλ-1 has so far not been manipulated successfully. In the receptor shown in Fig. 2 targets 3 and 5 have been removed by deletions leaving targets 2 and 4 spanning a sequence (fragment C) amounting to about 25% of the λ genome. Upon restriction of this DNA, fragment C can be replaced by a fragment (X) from a R.BamI digest of some other DNA and formation of a viable DNA molecule requires the correct rejoining of the four fragments A, B, X and D, which is inefficient. Fragments B, X and D (or B, C and D) however are joined together at a tolerable frequency. The left terminus, which is necessary for infectivity, may then be supplied in vivo by pre-infecting the cells with a helper phage that carries genetic markers (such as susJ and susR) to distinguish it from the DNA to be recovered. If the helper phage also has an immunity different from that of the phage to be recovered selection against the helper can be made either by plating the infected cells on indicator bacteria immune to the helper or by using cells lacking a suppressor for the amber mutations. In appropriate host cells recovery of the left cohesive end by recombination can be quite efficient and in this case recombinants of the desired type (i.e. with fragment X replacing fragment C) can be readily distinguished from parentals because they have acquired the Spi$^-$ phenotype, and will plate on a P2 lysogen (29). This selection has been used successfully with fragments of E.coli DNA inserted into the receptor, and the Spi$^-$ phage screened directly for specific genes by complementation of auxotrophs in the usual way (D. Ward and Noreen E. Murray, unpublished work).

Phage λ DNA may be packaged into virus particles in vitro (30, 31, 32,33) and with appropriate systems the process is very simple and quite efficient, yields as high as 0.1% having been attained although values around 10^{-4} (plaques per DNA molecule) are more common (32,34), and, of course, it avoids the variability found between strains since phage are recovered without propagation through a bacterial strain. The procedures make use of two lysogens (one of which is defective in genes for one essential head protein and the other in genes for another) which between them contain all the constituents necessary to convert λ DNA into a mature virus particle. In concentrated extracts of cells containing phage head precursor (which requires the products of phage genes E, B,C and Nu3) and the packaging proteins (products of phage genes A, D, Nu1 and F1) λ DNA is encapsidated and matured into plaque-forming particles in the presence of proteins from genes W and FII and phage tails (30,33). Of the two lysogens used, one is defective in gene E and the other in gene D so that neither can produce a mature phage itself, but accumulates the remainder of the precursors. On mixing extracts of both cells normal D and E proteins are present thus providing all of the components necessary for packaging the phage genomes. Endogenous, concatemeric phage DNA present in the lysates (from induction of the lysogens) is then packaged and cleaved (31), but if mature DNA from another lambdoid phage is added to the mixture this also is packaged and can be distinguished from the endogenous phage by use of a suitable immunity marker. If the two lysogens used to provide the components for phage assembly carry the immunity of phage 434 then phage resulting from in vitro recombination reactions with a receptor having a different immunity may be recovered selectively by plating the packaging reaction mixtures on E.coli lysogenised with λimm^{434} (34).

The efficiency of in vitro packaging is in general at least an order of magnitude, and often more, higher than that from transfection of the same DNA preparation (recombinant or otherwise) and a further practical advantage is that large batches of the packaging cells may be prepared and stored frozen (-57°C) for at least six months without any loss of competence for packaging (34).

With the in vitro packaging system described there was little variation in yield attributable to phage genotype nor was yield affected appreciably by deletions of various size within the normal range of genome size for viable phages (about 75 to 106% of λ^+) (31). To this extent the process appears to be non-selective and the whole population of viable recombinant DNA molecules is probably recovered. However, in some experiments with a mixture of a receptor genome (having a deletion of about 20%) and its in vitro recombinant a noticeably higher yield of the recombinant has been obtained by in vitro packaging, but not by transfection (B. Klein, B. Hohn and K. Murray, unpublished work).

In experiments with a somewhat different packaging system in which A^- and E^- lysogens were used, the efficiency was found to be markedly more efficient with genomes whose size approximated that of wild-type λ than with deletions and in fact provided a noticeable selection against the latter (36).

A further feature of the in vitro packaging systems is that recombination may occur in the reaction mixtures between the endogenous DNA and the DNA (from the in vitro recombination reactions) added to be packaged (34). Again the mechanism of this is obscure and in some cases the process may be desirable while in others it is clearly not. This recombination may be prevented (or its progeny rendered irrecoverable) by irradiation of the packaging cells with ultraviolet light (34), or by using as packaging cells lysogens so constructed that the replicated prophage DNA cannot be excised from the host chromosome (36).

Finally, since in vitro packaging does not involve propagation of phage genomes through cells it offers a degree of containment at the stage of recovery of the recombinant DNA molecules and this is enhanced by the chloroform treatment which kills bacterial cells and has been suggested as a means of chemical containment (37).

3. Selection and Screening of Recombinants

Many simple procedures are available to distinguish phage with recombinant genomes from those with the parent receptor. One of these which is very useful for selection of recombinants derived from prokaryotic DNA requires the functional expression of the inserted DNA fragment. Lytic propagation of the phage or growth of a transductant colony is made dependent upon the provision by the recombinant phage of a function in which the host strain is deficient (35,38). Two other convenient screening systems permit the easy recognition of recombinants by a change in phage phenotype (16).

One of these systems makes use of receptors that have a single restriction target located within the immunity region and have about 20% of the genome deleted. Insertion of any piece of DNA at this site makes it impossible for the phage to produce a functional repressor so that the recombinants give clear plaques which are readily distinguished from the turbid plaques of parental phage that result simply from rejoining of the two fragments of the receptor DNA molecule. Receptors of this type are available for use with the R.*Eco*RI and R.*Hin*dIII enzymes.

The second procedure is especially useful for cloning larger fragments of DNA and makes use of a phage DNA that has two targets for a particular restriction enzyme such that removal of the DNA (which contains non-essential functions) between these targets followed by joining of the two outer fragments would give a DNA molecule too small to yield an infectious phage. Production of a mature phage therefore requires insertion of additional DNA into the space between the two outer fragments of the receptor (Fig. 3).

Some recombinant phage made with *E.coli* DNA provide very good receptor systems because replacement of the *E.coli* DNA by other DNA fragments can be observed readily; these are phage containing suppressor genes, expression of which suppresses amber mutations in the host strain. An *E.coli lacZ* amber mutant is a convenient host for these phage because although the strain can grow without producing β-galactosidase simple colour tests are available to distinguish between the mutant and wild-type cells. On MacConkey indicator agar *E.coli* colonies (or *turbid* phage plaques) are red if they produce β-galactosidase and so hydrolyse lactose, but colourless if they do not. Phage that supply the suppressor function to relieve the amber mutation in *lac* are thus recognised as red plaques. An alternative colour test for β-galactosidase is based upon the hydrolysis of 5-bromo-

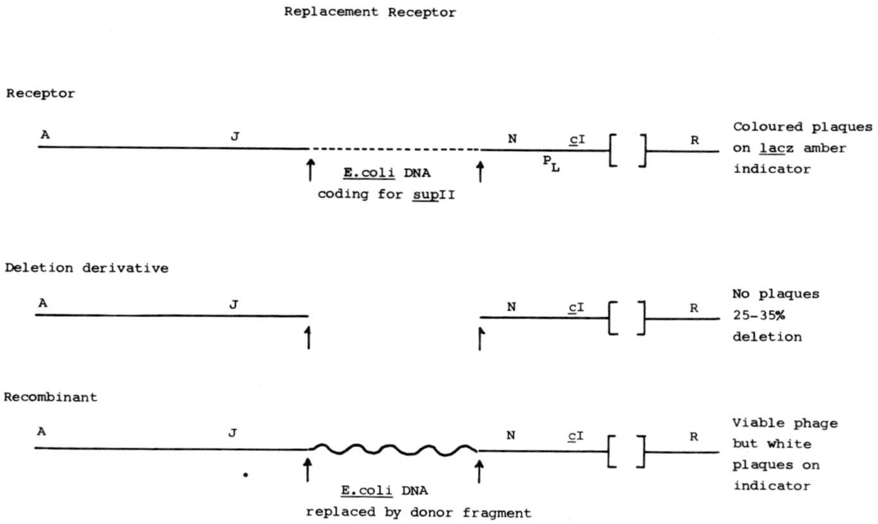

Fig. 3. Replacement receptors. The phage genome has two targets for the restriction endonuclease which span a segment of E.coli DNA carrying a suppressor gene. Removal of the central segment and joining of the two flanking fragments gives a DNA molecule too small for packaging. Replacement of the central segment by a fragment of DNA from another source permits the formation of viable phage, but these are no longer able to suppress an amber mutation (in a non-essential function) in the host cell.

4-chloro-3-indolyl-β-D-galactoside to release a non-diffusing blue pigment which gives blue plaques. This reagent may also be used with clear plaques whereas the MacConkey indicator requires that repression be established to give turbid plaques. When these phages are used as receptors (16), replacement of the DNA fragment between the two restriction targets by DNA from some other source gives rise to phage that cannot suppress the lac amber mutation in the host strain, so the recombinants are readily identified as colourless plaques.

Screening within recombinants for particular genes may be pursued in a variety of ways. Where a functional gene product is formed screening via complementation of a suitable auxotrophic host strain is convenient (38), but will not always be applicable. Another approach is to use a reaction that can be adapted to a colour test for the product of an enzyme function, such as those described for β-galactosidase; a further example is the decolouration of iodine by penicilloic acid produced from penicillin by a β-lactamase (39), by which lambda derivatives carrying penicillinase genes from Bacillus licheniformis have been detected (W. J. Brammar and S. Muir, unpublished observations). Immunological methods have also been used to demonstrate the formation of β-galactosidase from lac transducing derivative of λ and these powerful techniques are clearly of wide application and offer the possibility of detecting expression of the inserted DNA segments to produce proteins that may be functionally defective (40,41). When a suitable radioactive probe is available, in situ hybridisation methods offer powerful and rapid screening procedures. These are especially useful in the analysis of recombinants carrying eukaryotic DNA and are described in Section 5.

4. Expression of Genes Inserted into Lambda

The replacement receptors and screening of recombinants via complementation or some of the other procedures described depend for their use upon the correct expression of genes acquired by the recombinants. Expression of these genes may be from their own promoters or from a phage promoter, but in either case the phage regulatory systems can be exploited to enhance the level of expression of the inserted genes (21,42). Some of the essential features of lambda are summarised in a very simple form in Fig. 4 and its legend.

The products of genes S and R are necessary for lysis of the host cell; an amber mutation in gene S prevents cell lysis but allows DNA replication and protein synthesis to continue. In this way the number of gene copies is greatly increased and their products accumulate in the cell, as was demonstrated for the lac repressor (44). Cell lysis may also be prevented by an amber mutation in the regulatory gene, Q, which controls expression not only of gene S, but also of the late genes located on the left of the linear map (see Fig. 4). The proportion of the cell's protein synthesising activity directed through an incorporated promoter is thus increased. By further genetic manipulation of the phage efficient expression of genes inserted, for example at srIλ-3 or shnλ-3, may be achieved from the phage promoter, P_L. This expression may also be aided by the lambda N gene product which interacts with RNA polymerase enabling it to override transsciptional stops (45). With systems of this sort, the yield of protein can be very high, certainly greatly exceeding that from a normal bacterial cell. An example is provided by the admittedly favourable case of the trp operon of E.coli where derivatives of transducing phages (made in vivo by conventional

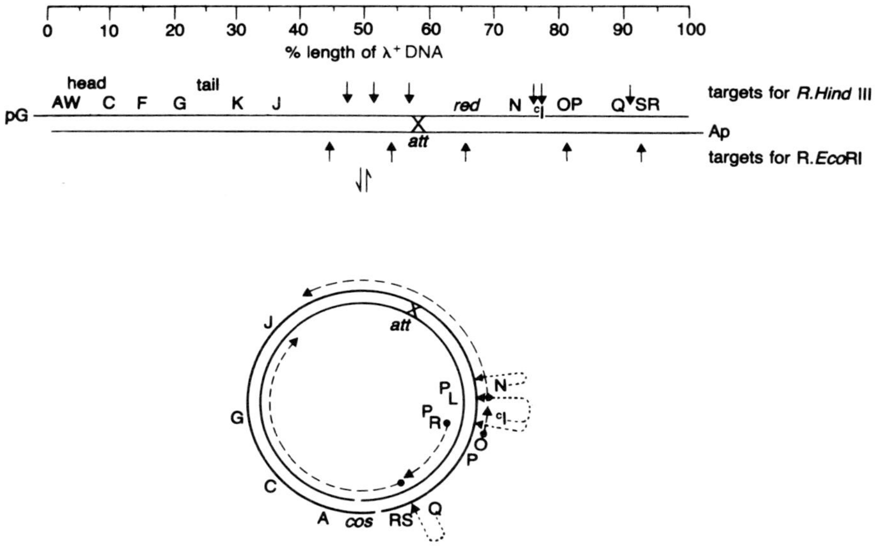

Fig. 4. *A simplified version of the lambda chromosome showing some of its more important control functions. The mature phage contains a single, linear duplex DNA molecule with a molecular weight of 3.1 x 10^7. At the 5' ends of the DNA are single-stranded projections of 12 bases with complementary sequences. Upon injection into the host cell, the DNA circularises by base pairing of these cohesive ends. Chromosomes of lambdoid phages are normally drawn in the linear form; here genes are located at the positions corresponding to the percentage of the length of the wild-type phage DNA. Genes on the left of the linear map code for head and tail proteins of the phage. Much of the central region is inessential and can be deleted without seriously impairing phage growth. Red represents the phage recombination system, O and P are concerned with replication of the phage DNA, and S and R code for proteins that lyse the host cell when the phage products have been assembled into infectious particles. The cI gene codes for a repressor protein which interacts at the sites shown by the dotted arrows in the lower part of the figure to prevent expression of the phage genes. Removal of the repressor permits expression in both directions from P_L and P_R as shown by the broken arrows. N and Q are positive regulatory genes, the products of which interact at the positions shown by dotted arrows. Q is necessary for the expression of genes S and R and genes to the left of these as indicated by the long broken arrow inside the circle. Thus, after circularisation of the chromosome, gene Q activates the expression of genes A, C etc. (that is, those on the left of the linear map) as well as genes R and S.*

Also shown in the figure is the attachment site by which the phage chromosome may be inserted into its host chromosome (where it may be stably replicated along with the host), and the positions of targets for the restriction enzymes R.HindIII and R.EcoRI in the wild-type chromosome. Some of these targets must be removed by deletion or mutation in order to make phage derivatives that can be used as receptors (see Fig. 1)

genetic methods) have been produced which make over 50% of their soluble protein as products of these genes (46). Operations of this sort are being used to improve the production of a number of enzymes of importance in nucleic acid biochemistry, such as the DNA ligase of E.coli (47), DNA polymerase, and DNA restriction and modification enzymes (N. E. Murray and K. Murray, unpublished work). Various ways in which the recombinant phages may be manipulated for these purposes are described in refs. 42, 43, 46.

5. Eukaryotic DNA Inserted into Lambda

The various lambda receptor systems can obviously be used for any DNA fragments regardless of their provenance and some of the replacement systems are particularly suited to larger DNA fragments. For DNA fragments ranging up to about 6.10^6 M.W. the immunity insertion systems offer a very simple distinction between recombinant and parental phage (16).

Selection of a given recombinant of a eukaryotic sequence is complicated by the complexity of eukaryotic DNA so that some sequences will be expected in low abundance. Further, the metabolic pathways of many eukaryotic organisms differ appreciably from those of E.coli so that although some encouraging results have been obtained with complementation of E.coli auxotrophic strains by hybrids of lambda (and plasmids, 48) carrying yeast DNA (47,49), this will frequently be unsatisfactory as a screening method.

Where suitable probes are available, such as purified messenger RNA preparations, nucleic acid hybridization methods offer an attractive and obvious approach. The in situ methods used so successfully in cytological analyses (50) were shown to be applicable to individual phage plaques in test experiments with phage that carried two targets (srIλ-1 and srIλ-2) for R.EcoI (51). The DNA fragment contained between these two targets was recovered after restriction of the DNA with R-EcoRI and used as a template for the synthesis of cRNA with E.coli RNA polymerase and radioactively labelled nucleoside triphosphates (Fig. 5). This labelled RNA served as a probe for phage carrying its complementary sequence and in the experiment illustrated in Fig. 5 about 2,700 cpm of radioactivity was retained by single plaques of phage carrying the central DNA fragment, while controls with phage from which this segment was deleted, or with plaques of phage T4 or equivalent areas of the bacterial lawn gave essentially zero hybridization. From this result one would expect fragments about 1% of the size of the lambda genome (i.e. M.W. about 3.10^5, or about 400 base pairs) to be detectable with confidence if the specific radioactivity of the probe were about 10^7 cpm per μg and its

Detection of DNA sequences by in situ hybridisation in phage plaques.

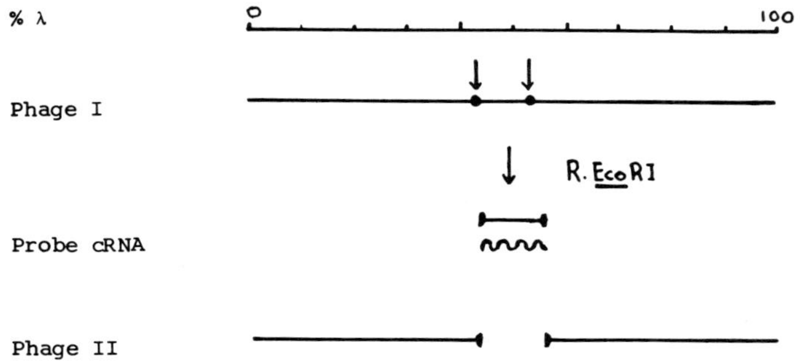

Cts/min retained after hybridisation with labelled cRNA probe:

In plaques		Control sample
Phage I	Phage II	from lawn
2698	36	21

Fig. 5. In situ hybridization on individual phage plaques. The genome of the phage represented in the upper part of the figure contains two targets for the EcoRI enzyme. A transcript of the fragment contained within these two targets was made in vitro with RNA polymerase and radioactively labelled nucleoside triphosphates. This RNA was used as a probe to test the efficiency of hybridization on individual plaques of phage containing these sequences, the phage with this segment deleted, as well as T4 phage and a section of the bacterial lawn serving as controls.

size similar to that of the inserted fragment. The *in situ* hybridization method has now been developed and improved considerably and the elegant version of it described in this volume permits very large numbers of phage to be screened rapidly and economically (47).

Recombinant phage carrying sea urchin DNA were screened for the presence of histone genes by a modification of the earlier *in situ* hybridization method with histone mRNA. The sea urchin DNA used in these experiments had been digested with R.*Eco*RI and fractionated by electrophoresis in agarose gels from which the fragments were eluted and those containing histone gene sequences were located by hybridization with mRNA. It was necessary to purify the eluted fragments, by equilibrium centrifugation in caesium chloride solution, before insertion into the phage genome otherwise the ligase reaction was completely inhibited. An immunity insertion receptor (16) was used and from 200 recombinants screened (clear plaques), 3 gave positive results on hybridization with histone mRNA. These phage were propagated and analysis of their DNA by digestion with several restriction enzymes showed that two contained the sea urchin fragment in one orientation and one in the other; this is illustrated schematically at the top of Fig. 6. The two strands of λ DNA are readily separated by equilibrium centrifugation in caesium chloride solution in the presence of poly rUG (52). When the strands of DNA from two of the recombinants containing histone genes in opposite orientations were separated in this way the histone mRNA, which was a mixture of all five classes of histone mRNA, hybridised uniquely against one strand - the heavy strand in one case and the light strand in the other, as shown in Fig 6 - which means that all of the histone genes are present on the same strand of the DNA (P. Mounts, E. M. Southern and K. Murray, unpublished work). This result has also been obtained from similar experiments, but with the DNA strands separated by electrophoresis in agarose gels, with sea urchin DNA digested with R.*Hin*dIII (53) and cloned in the appropriate immunity insertion vector (16), and with sea urchin histone genes cloned in a plasmid (54).

Histone genes offer an interesting example for studies of the expression of eukaryotic genes in bacteria since bacteria do not normally make these proteins or have the need of them. Recombinants of these genes in λ are attractive for this purpose since transcription of the inserted genes from a phage promoter can be made conditional through the use of appropriate regulatory mutants (42,43). However, transcription of genes inserted in the immunity receptors would be dependent upon the use of their own promotors which may be

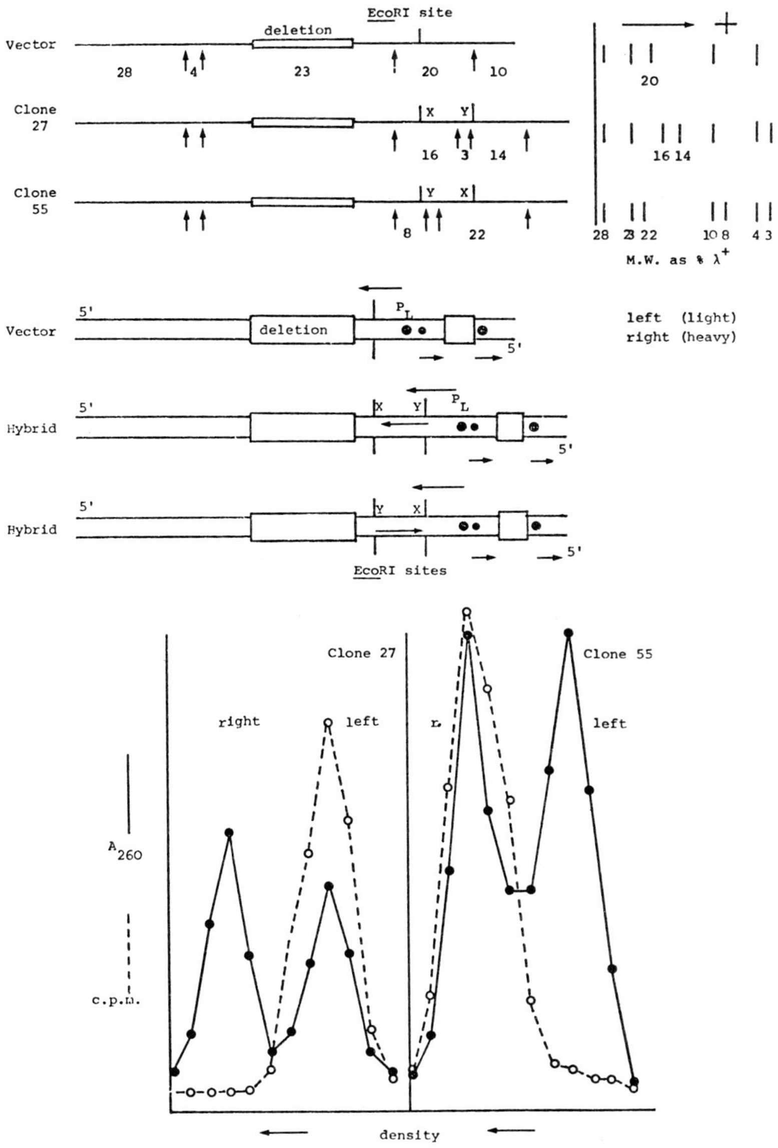

Fig. 6. Analysis of DNA from hybrids of lambda and sea urchin DNA. The upper part of the figure illustrates ***schematically*** *the distinction between two recombinants with the DNA fragment in opposite orientations by electrophoretic analysis on agarose gels of a digest of the DNA from the two phages (and their parent phage) with a second restriction enzyme. The vertical arrows show targets for this restriction enzyme and the numbers between them (and beneath the bands separated by*

gel electrophoresis as shown on the upper right of the figure) give the size of the DNA fragments as % of λ^+ DNA. An immunity insertion receptor was used for cloning the sea urchin DNA, but for expression of the inserted genes it is desirable (perhaps necessary) to transfer the DNA fragment to a different receptor where expression is regulated by the phage promoter, P_L. Two orientations are again possible, the desirable one having the direction of transcription from right to left, as shown in the central part of the figure; the horizontal arrows indicate directions of transcription from the more important promoters which are located by the dots. This may be determined by separation of the two strands of the phage DNA by equilibrium centrifugation followed by hybridization of fractions from the CsCl gradient with radioactive mRNA as illustrated in the lower part of the figure for DNA from two of the original clones in the immunity insertion receptor.

inefficient, particularly with eukaryotic sequences, and it is therefore desirable to transfer the fragment to another receptor phage where, when inserted in the correct orientation, its expression will be under control of the efficient λ promoter, P_L. When sea urchin (Echinus esculentus) histone genes that had been cloned in an immunity receptor for R.EcoRI digests of DNA (in which they are readily propagated and stable) were transferred in vitro to such a receptor, in which they were inserted at srIλ-3, the recombinants proved to be unstable and the fragment was rapidly and consistently lost (K. Peden, B. Klein and K. Murray, unpublished work). Further manipulation of the inserted fragment is now proceeding to establish a stable recombinant for studies of expression.

The 2μ circular DNA from yeast provides a further example where studies of a eukaryotic DNA have been pursued via in vitro recombination with a suitable λ receptor. As isolated from yeast cells this DNA is always obtained as a mixture of two forms, due, it is believed, to the presence of an inverted repeated sequence (55). When cloned in λ DNA a homogeneous single form is stable and can be prepared in quantity for further structural and mapping studies (56) and this species is the focus of much current attention as a possible receptor molecule for use with eukaryotic cells.

This brief survey has been intended to provide some examples of the way in which bacteriophage λ can be exploited in various aspects of research on recombinant DNA molecules made in vitro. To these examples should be added the ease with which heteroduplex electron microscopy and denaturation mapping can be pursued with lambda and its derivatives (1, 57).

ACKNOWLEDGMENTS

I am grateful to various colleagues mentioned in the text for permission to refer to their unpublished work. The genetic manipulations of λ which I have described were almost entirely the work of my wife, Noreen E. Murray. My research described here and elsewhere has been greatly helped by the loyal assistance of Sandra Bruce and has been supported by the Science Research Council. Figs. 2-6 are reproduced with permission.

REFERENCES

(1) A. D. Hershey (editor), The Bacteriophage Lambda (Cold Spring Harbor Laboratory, N.Y. 1971).

(2) N. E. Murray and K. Murray, Nature Lond. 251, (1974) 476.

(3) H. O. Smith and D. Nathans, J. Mol. Biol. 81, (1973) 419.

(4) R. J. Roberts, CRC Crit. Revs. Biochem. 4, (1976) 123.

(5) K. Murray, Proc. Roy. Soc. (1977) In press.

(6) B. Allet, P. G. N. Jeppesen, K. J. Katagiri and H. Delius, Nature New Biol. 241 (1973) 120.

(7) K. Murray and N. E. Murray, J. Mol. Biol. 98 (1975) 551.

(8) T. Hohn, Phil. Trans. Roy. Soc. B, 276 (1976) 143.

(9) A. J. D. Bellet, H. G. Busse and R. L. Baldwin, in The Bacteriophage Lambda, ed. A. D. Hershey (Cold Spring Harbor Laboratory, N.Y. 1971) p. 501

J. Weil, R. Cunningham, R. Martin III, E. Mitchell and B. Bolling, Virology, 50 (1973) 373.

(10) K. J. Danna, G. Sack and D. Nathans, J. Mol. Biol. 78 (1973) 363.

(11) P. A. Sharp, W. Sudgen and J. Sambrook, Biochemistry 12 (1973) 3055.

(12) M. Thomas and R. W. Davis, J. Mol. Biol. 91 (1975) 315.

(13) P. H. Seeburg and H. Schaller, J. Mol. Biol. 92 (1975) 261.

(14) H. O. Smith and M. L. Birnstiel, Nucleic Acids Res. 3 (1976) 2387

(15) W. Arber and S. Linn, Ann. Rev. Biochem. 38 (1969) 467.

(16) N. E. Murray, W. J. Brammar and K. Murray, Molec. Gen. Genetics 150 (1977) 53.

(17) W. Arber and V. Kuhnlein, Pathol. Microbiol. 30 (1967) 946.

(18) N. E. Murray, P. Manduca de Ritis and L. A. Foster, Molec. Gen Genetics 120 (1973) 261.

(19) M. Thomas, J. R. Cameron and R. W. Davis, Proc. Natl. Acad. Sci. U SA 71 (1974) 4579.

(20) A. Rambach and P. Tiollais, Proc. Natl. Acad. Sci. USA 71 (1974) 3927.

(21) N. E. Murray, in: Genetic Engineering, Ed. A. Chakrabarty (CRC Publishing Co., Cleveland, Ohio, 1977) In press.

(22) C. P. Bahl, K. J. Marians, R. Wu, J. Stawinsky and S. A. Narang, Gene 1, (1977) 81.

(23) M. Mandel and A. Higa, J. Mol. Biol. 53 (1970) 159.

(24) E. M. Lederberg and S. N. Cohen, J. Bact. 119 (1974) 1072.

(25) A. D. Kaiser and D. S. Hogness, J. Mol. Biol. 2 (1960) 392.

(26) G. A. Wilson and F. E. Young, J. Mol. Biol. 97 (1975) 123.

(27) D. M. Haggerty and R. F. Schlief, J. Virol. 18 (1976) 659.

(28) M. Perricaudet and P. Tiollais, FEBS Letters 56 (1975) 7.

(29) J. Zissler, E. Signer and F. Schaefer, in: The Bacteriophage Lambda, ed. A. D. Hershey (Cold Spring Harbor Laboratory, N.Y. 1971) p. 469.

(30) S. Casjens, T. Hohn and A. D. Kaiser, J. Mol. Biol. 64 (1972) 551.

(31) B. Hohn and T. Hohn, Proc. Natl. Acad. Sci. USA, 71 (1974) 2372.

(32) B. Hohn, J. Mol. Biol. 98 (1975) 93.

(33) T. Hohn, M. Wurtz and B. Hohn, Phil. Trans. Roy. Soc. Lond. B. 276 (1976) 51.

(34) B. Hohn and K. Murray, Proc. Natl. Acad. Sci. USA (1977) In press.

(35) J. R. Cameron, S. M. Panasenko, I. R. Lehman and R. W.Davis, Proc. Natl. Acad. Sci. USA 72 (1975) 3416.

(36) N. Sternberg, D. Tiemeier and L. Enquist, Gene, 1, (1977) In press.

(37) C. Weissmann and W. Boll, Nature Lond. 261 (1976) 428.

(38) K. Borck, J. D. Beggs, W. J. Brammar, A. S. Hopkins and N. E. Murray, Molec. Gen. Genetics 146 (1976) 199.

(39) D. J. Sherratt and J. F. Collins, J. Gen. Microbiol. 76, (1973) 217.

(40) B. Sanzey, O. Mercereau, T. Ternynzk, and P. Kourilsky, Proc. Natl. Acad. Sci. USA. 73 (1976) 3394.

(41) A. Skalka and L. Shapiro, Gene 1 (1977) 65.

(42) N. E. Murray, in: Recombinant DNA: Proceedings Xth Miles International Symposium. (Raven Press Inc. 1977) In press.

(44) B. Müller-Hill, L. Crapo and W. Gilbert, Proc. Natl. Acad. Sci. USA. 59 (1968) 1259.

(45) N. C. Franklin, J. Mol. Biol. 89 (1974) 33.

(46) A. Moir and W. J. Brammar, Molec. Gen. Genetics 149 (1976) 87.

(47) R. W. Davis, this volume.

(48) J. Carbon, this volume.

(49) K. Struhl, J. R. Cameron and R. W. Davis, Proc. Natl. Acad. Sci. USA. 73 (1976) 1471.

(50) H. John, M. L. Birnstiel and K. W. Jones, Nature Lond. 223 (1969) 582.

(51) K. W. Jones and K. Murray, J. Mol. Biol. 96 (1975) 455.

(52) W. Szybalski, H. Kubinski, Z. Hradecna and W. C. Summers, in: Methods in Enzymology, eds. L. Grossman and K. Moldave, (Academic Press Inc., N.Y. Vol. XXID (1971) p. 383.

(53) K. Gross, W. Schaffner, J. Telford and M. L. Birnstiel, Cell 8 (1976) 479.

(54) L. H. Keddes, in: Recombinant DNA: Proceedings Xth Miles International Symposium (Raven Press Inc. 1977) In press.

(55) M. Guerineau, C. Grandchamp and P. P. Slonimski, Proc. Natl. Acad. Sci. USA. 73 (1976) 3030.

(56) J. D. Beggs, M. Guerineau and J. F. Atkins, Molec. Gen. Genetics 148 (1976) 287.

(57) R. Portmann, W. Schaffner and M. L. Birnstiel, Nature Lond. 264 (1976) 31.

DISCUSSION

R.B. MEAGHER: Using the plasmid vectors for molecular cloning, it has been difficult to clone DNA fragments with molecular weights greater than about 15,000,000 because of the inefficiencies in transformation, and you have vehicles here that would accommodate 10,000,000 daltons of DNA. Do you think it would be possible to design a lambda vehicle that would accommodate a much larger fragment of DNA, of say 25,000,000 daltons?

K. MURRAY: One can package longer lambda chromosomes containing up to about 105%, or even 109% of wild type, DNA. These phages, however, tend to be unstable and to accumulate deletions. I would think that 15,000,000 should be quite possible with a bit of work; Fig. 1 shows that about 43% of the lambda genome can be replaced in a suitable phage.

R.B. MEAGHER: What about the whole way - you implied that you could go to even larger amounts of DNA in theory - so why only 43%, why not 75%?

K. MURRAY: Such genomes would have to be handled as defective phages. Those that I have been talking about can be handled quite easily. Handling defective phages is more troublesome.

B. WEISBLUM: Could it be possible that unstable ones are those that make histones.

K. MURRAY: Yes, this is possible, but I doubt that it is the complete explanation for our failure to transfer the fragments. I would imagine that the fragment will be inserted in both orientations, and I expect it to be expressed in only one orientation.

RAPID SELECTION AND SCREENING METHODS FOR THE ISOLATION OF PARTICULAR CLONED EUKARYOTIC DNA SEQUENCES

R. W. DAVIS, M. THOMAS, D. BENTON, J. CAMERON, P. PHILIPPSEN, K. STRUHL, T. ST. JOHN, R. KRAMER
Department of Biochemistry
Stanford University School of Medicine
Stanford, California 94305

Abstract: Genetic and physical methods have been developed which allow one to rapidly isolate a desired eukaryotic DNA sequence cloned in a λgt vector. The genetic selection is based on complementation of nonrevertible *Escherichia coli* mutations. A sequence coding for an imidazole glycerol phosphate dehydratase has been isolated from *Saccharomyces cerevisiae* by such a selection. Using ^{32}P complementary RNA or DNA, a new plaque hybridization method has been developed that allows one to screen up to 10^6 plaques per day for a particular sequence. By using *pel*⁻ host strains and a particular λgt vector, one can select the size of the cloned DNA. A method has also been developed that allows one to directly examine the restriction spectrum from a crude agarose plate lysate.

INTRODUCTION

Recently it has become rather simple to clone DNA sequences using *E. coli* vectors. If one has a homogeneous DNA segment to be cloned, it is also rather simple to find the appropriate chimeric molecule among the limited variety of joined molecules. However, if the desired segment is but one among a number of segments, then finding the appropriate chimeric molecule can often be a laborious task. This report describes a number of selection and screening methods that allow one to rapidly isolate the appropriate chimeric molecule even in the presence of large numbers of inappropriate molecules.

EXPERIMENTAL

The methods for genetic selection have already been described (1). Plaque hybridizations are conducted by first transferring the DNA in plaques to a nitrocellulose membrane filter, denaturing the DNA in NaOH, drying the filter, and then adding the radioactive complementary nucleic acid. The DNA is transferred to the filters by making direct contact with a dry filter to an agar plate with up to 2×10^4 plaques for about 5 min. The agarose plate lysates are made by overlaying a confluently lysed plate with 5 ml cold λ diluent (10 mM Tris-HCl, pH 7.5, 10 mM $MgSO_4$) overnight at 5°C. To 0.4 ml of this in a microfuge tube is added 40 μl 0.5 M Na_3EDTA, 20 μl 2 M Tris base and 20 μl 10% sodium dodecyl sulfate. After mixing, 1 μl diethyloxydiformate is added with shaking and then heated to 65° for 30 min. The tubes are cooled in ice and 0.1 ml 5 M potassium acetate is added. After one hour at 0°, the precipitate is sedimented in an Eppendorf centrifuge 3200 for 10 min. The nucleic acid in the supernatant is precipitated by the addition of 1.1 ml ethanol. After 1 hr at -20°, the precipitate is sedimented and redissolved in 40 μl of 0.01 M Tris-HCl, pH 7.5, 10^{-3} M Na_3EDTA. Five microliters is usually sufficient for gel electrophoresis.

RESULTS AND DISCUSSION

Genetic selection. One of the simplest methods for isolating a eukaryotic sequence of known function is by complementation of known *E. coli* mutations. It has already been shown that such complementation can occur (1). λgt-Sc2601 contains a 10 kilobase (kb) DNA segment from *S. cerevisiae* (baker's yeast) that, when integrated into an *E. coli his* B mutant lacking imidazole glycerol phosphate (IGP) dehydratase, allows this bacterium to grow in the absence of histidine. IGP dehydratase activity is found in cell extracts when the yeast sequence is integrated into the mutant chromosome. This is not due to suppression of the original mutation because the enzyme can also be found in a cell that is deleted for the entire *his* operon.

Plaque hybridization. Methods have already been developed for screening colonies (2) or plaques (3, 4) for

DNA sequences complementary to a radioactive nucleic acid probe. We have greatly simplified these methods thus allowing larger numbers of clones to be screened. Free DNA and phage in a plaque are rapidly adsorbed when contact is made to a dry nitrocellulose filter. Therefore, by simply dropping an untreated dry filter (millipore HA) onto an agar plate containing up to 2 x 10^4 plaques, a replica is produced suitable for hybridization. Some of the DNA and phage pass through the filter thus allowing stacks of filters to be prepared in one operation which provides up to 5 identical hybridizable filters. Approximately 10^6 plaques can be screened per day.

Size selection. Bacteriophage lambda containing a particular size of DNA can be selected physically by its buoyant density (5) or genetically by growth on a pel⁻ host (6, 7). The pel selection is very stringent for phage containing DNA of wild type size or larger. There are a number of λgt cloning vectors available with varying amounts of DNA in the vector portion. This allows one to select particular sizes of inserted DNA after cloning (Table 1). The efficiency of this selection is illustrated in Table 2.

TABLE 1

Vector	Joints	cI [1]	red [2]	DNA size to w.t [3]
λgt 1	EcoRI	ts	−	13 kb
λgt 2	EcoRI	ts	+	7.8 kb
λgt 3	EcoRI	ts	−	8.3 kb
λgt 4	EcoRI	ts	+	10 kb
λgt 5	EcoRI	ts	−	15 kb
λgt 6	EcoRI	ts	+	13 kb
λgt 7	EcoRI	del	+	14 kb
λgt 20	HindIII	del	−	19 kb
λgt 21	HindIII	del	+	9.2 kb
λgt 40	SstI	ts	−	9.7 kb

1) cI = lambda repressor gene. ts is the cI857 mutation and del is the HindIII·E fragment deleted.
2) red = lambda recombination. red^- phage give a yield of about 1/3 that of red^+ phage.
3) The size of the DNA in kilobase pairs (kb) that restores the vector to wild type in size. Overpackaged phage up to 5 kb additional DNA will also be selected by the pel⁻ strains.

TABLE 2

Efficiency of plating of λgt 2 containing various lengths of inserted DNA on pel⁻ host

Size of insert	Efficiency of plating
0	10^{-4}
0-5 kb	10^{-2}
5-8 kb	1

Restriction spectra from agarose plate lysates. Frequently the best method for identification of a desired chimeric molecule is by its restriction spectrum (pattern of bands produced on gel electrophoresis of restriction endonuclease cleaved DNA). This approach becomes laborious when each phage to be examined must be purified by CsCl density gradient centrifugation and the DNA prepared by phenol extraction. We have developed a simple and rapid method which is based on the observation that about half of the DNA released in plate lysates is phage DNA. The restriction spectrum of this DNA shows predominately the phage DNA bands in a background of faint E. coli DNA bands. Agarose is used because agar apparently contains potent inhibitors of some restriction endonucleases. The liquid overlay from a single plate lysate is treated with diethyloxydiformate and SDS and the protein precipitated with potassium acetate followed by ethanol precipitation of the DNA. Approximately 100 DNA samples can easily be prepared per day.

REFERENCES

(1) K. Struhl, J. R. Cameron and R. W. Davis, Proc. Nat. Acad. Sci. USA, 73 (1976) 1471.

(2) M. Grunstein and D. Hogness, Proc. Nat. Acad. Sci. USA, 72 (1975) 3961.

(3) K. Jones and K. Murray, J. Mol. Biol., 96 (1975) 455.

(4) R. A. Kramer, J. R. Cameron and R. W. Davis, Cell, 8 (1976) 227.

(5) G. Kellenberger, M. L. Zichichi and J. Weigle, J. Mol. Biol., 3 (1961) 399.

(6) S. W. Emmons, V. MacCosham and R. I. Baldwin, J. Mol. Biol., 91 (1975) 133.

(7) D. Scandella and W. Arber, Virology, 69 (1976) 206.

We wish to thank Lynn Horn for her help in preparing this manuscript. This work was supported in part by Public Health Service Grant GM 21891 from the National Institutes of General Medical Sciences.

DISCUSSION

P. DUESBERG: What is the evidence that this dehydrogenase is indeed a yeast enzyme? Do you have fingerprint of tryptic digests?

R.W. DAVIS: You'll never prove that you have the same enzyme made in yeast by doing enzymology simply because yeast may have many gene copies, some of which might be silent. We may have cloned a silent gene which is not expressed under any laboratory conditions. If you try to identify it by tryptic fingerprints you may conclude it is identical but it will not be the gene that is expressed in yeast. So I think the only way of approaching this is through cloning of mutants. The evidence is that *his3* is the yeast dehydratase gene. However, *his3* may not be the dehydratase structural gene but a control gene. Possibly the best way of identifying the structural gene of yeast dehydratase is by cloning it in *E. coli*. If we find, in fact, that the cloned mutants of *his3* do not function in *E. coli* and if we can clone a suppressible mutant and suppress it in *E. coli*, we would have the best evidence for expression of the yeast gene. If we can clone a number of mutants, which is what we intend to do, we can map these mutations by deletion analysis exactly the way that the *E. coli* trp operon has been mapped. That is why we are making a whole set of deletions in *his*. We can also reconstruct the functional gene by recombination of yeast DNA in *E. coli*. That is the kind of approach that has been used in *E. coli*.

W. SZYBALSKI: Some people might object that your experiments with cloning DNA from *his3* are negative type of evidence.

R.W. DAVIS: We intend to clone a number of different mutants, that will recombine in yeast to give a functional gene. We will simply take those clones, recombine them in *E. coli* and show that we can re-establish a functional gene. That will not be a negative result.

B. MACH: You could combine the R-loop formation with the binding protein and antibody technique as a powerful gene enrichment procedure. . .

R.W. DAVIS: We haven't really considered that because the major contaminant that remains after making the R-loops is single stranded DNA. If there are two nicks close together when you make R-loops, you melt out the small region in between and you obtain DNA with a small gap which will also bind the single-strand specific protein. We have done some enrichment using B.D. cellulose columns which bind single stranded DNA.

W. SZYBALSKI: Is it really necessary to go through the rather involved procedure to distinguish single-stranded DNA from the RNA-DNA hybrid? In our hands we find that there is a clear difference using uranyl-oxide staining instead of platinum-palladium.

R.W. DAVIS: You can map large ribosomal RNAs that way but it is very difficult to see something the size of tRNA.

W. SZYBALSKI: You showed that one end of ribosomal RNA of yeast is not annealed perfectly and therefore appear as a whisker. We find the same situation with *E. coli* ribosomal RNA. We cannot explain this behavior in terms of the published sequence.

BACTERIAL PLASMIDS CONTAINING SILK GENE SEQUENCES

JOHN F. MORROW, NANCY T. CHANG,
JOHN M. WOZNEY, ANNE C. RICHARDS
and ARGIRIS EFSTRATIADIS*
Department of Biological Chemistry
Harvard Medical School
Boston, Massachusetts 02115
and
*Biological Laboratories
Harvard University
Cambridge, Massachusetts 02138

Abstract: A recombinant plasmid containing about 1200 base pairs of fibroin gene sequence (a repetitive, satellite-like DNA) is rather stable during many generations of growth in bacteria. Rare deletions have been found, however. For isolation of genes with their neighboring DNA sequences from large genomes, the use of messenger RNA linked to cellulose is described.

INTRODUCTION

We are using recombinant DNA techniques to study a eukaryotic gene encoding a differentiated protein product, silk fibroin of the commercial silk moth, *Bombyx mori*. We have prepared bacterial recombinant plasmids containing portions of the fibroin gene sequence (1) by the approach taken previously for construction of rabbit β-globin gene plasmids, the first step of which is reverse transcription of the messenger RNA (2-4). The assumption underlying the use of cloning in bacteria for study of eukaryotic genes is that the DNA introduced into bacteria is identical in sequence to the corresponding DNA obtained from them after many generations of growth. We report here experiments testing the fidelity of cloning of fibroin gene sequences.

The DNA adjacent to the fibroin gene is of interest because of its presumed role in regulation of the gene's transcription. We report here preliminary studies on a method which should be widely applicable for gene purification. It involves use of messenger RNA to purify the corresponding gene with its neighboring DNA from the total DNA of the organism (in this case, the silk moth).

EXPERIMENTAL

Fibroin messenger RNA was purified on the basis of its large size (5) from Bombyx mori Japanese inbred strain p22. E. coli K12 strain HB101, which is hsr^-recA^-, was obtained from H. W. Boyer. It was used as the host for recombinant plasmids, which were derived from plasmid pMB9 (6).

Fibroin plasmids were described previously (1). They were prepared by a method established for globin plasmids (4). It involves reverse transcription of fibroin mRNA, synthesis of the second DNA strand by DNA polymerase I, cleavage of the hairpin loop by S_1 nuclease, addition of homopolymer T tails by terminal transferase, annealing with pMB9 DNA with A tails added at the EcoRI site, and transformation of E. coli.

Fibroin mRNA was labeled with ^{125}I in vitro (7). DNA-RNA hybridization was performed as described by Suzuki et al. (8), in 50% formamide, 0.6 M NaCl, 0.12 M Tris Cl, pH 8, 8 mM EDTA, with DNA bound to nitrocellulose filters (Millipore Co., HA type). RNA was in excess. The temperature used was that found to give maximal hybridization rate in a preliminary experiment: 50° for fibroin mRNA with pBF39 DNA, 55° for ribosomal RNA of Bacillus subtilis or Bombyx mori with homologous DNA. Hybridization vials were agitated at 200 rpm in a shaker bath during incubation.

Finely divided cellulose was prepared (9), activated with 10 mg./ml. CNBr, and coupled to fibroin mRNA (10). SEP is 0.6 M NaCl, 8 mM EDTA, 0.12 M sodium phosphate, pH 6.0. The mRNA-cellulose was washed at 55° in SEP containing 50% formamide. 90% of the mRNA was coupled to the cellulose. It remained bound during 16 hr. incubation at 55° or months of storage at 4°. DNA was annealed with mRNA-cellulose in SEP-50% formamide at 55° with agitation. The concentration of mRNA was 0.15 µg./ml. Annealing was terminated by diluting with an equal volume of SEP, chilling to 0°, and centrifuging at 8000 rpm for 2 min. The mRNA-cellulose pellet was then washed in SEP and suspended in Aquasol-2 (New England Nuclear Co.) for liquid scintillation counting. In experiments using unfractionated Bombyx mori DNA, the mRNA-cellulose was washed in SEP-80% formamide at 25° after annealing. After three washes, less than 0.1% of the total Bombyx DNA (^3H-labeled) remains on the mRNA-cellulose. DNA can be eluted by incubation in 0.1 N NaOH at 40° for 20 min., or in 20 µg./ml. pancreatic ribonuclease, 0.2 M NaCl, 0.01 M Tris, pH 8 at 37° for 30 min., followed by washing in SEP-80% formamide.

Electrophoresis of DNA fragments was done in 1.4% agarose gels (11).

RESULTS

Recombinant plasmids containing fibroin gene sequences were obtained by a modification of a method previously established for globin genes (1-4). The purified plasmid DNAs were initially compared to one another by hybridization with fibroin mRNA in excess over the plasmid DNA bound to a nitrocellulose filter. The one which hybridized most rapidly with fibroin mRNA was pBF39.

The plasmid pBF39 hybridizes with the major, repeating sequence of fibroin mRNA. One μg of pBF39 DNA hybridized 87% of 8 ng of fibroin mRNA in a 12 hour incubation, and an identical, fresh pBF39 filter hybridized 2.4% in a subsequent incubation, so that 89.4% of the RNA was hybridized in all. The portion of pBF39 derived from synthetic duplex DNA is about 1100 base pairs, a small fraction of the length of fibroin mRNA, 16,000 bases (12). Consequently the mRNA sequence hybridizing to pBF 39 must be repetitious. Most of the fibroin polypeptide chain consists of about 700 repeats of the sequence gly-ala-gly-ala-gly-ser (13), and the predominant sequences in the mRNA are consistent with this (14).

We studied the rate of hybridization of pBF39 DNA with fibroin mRNA present in excess, since the rate is inversely proportional to the complexity of the RNA (15, 16). This annealing reaction is very fast (Fig. 1).

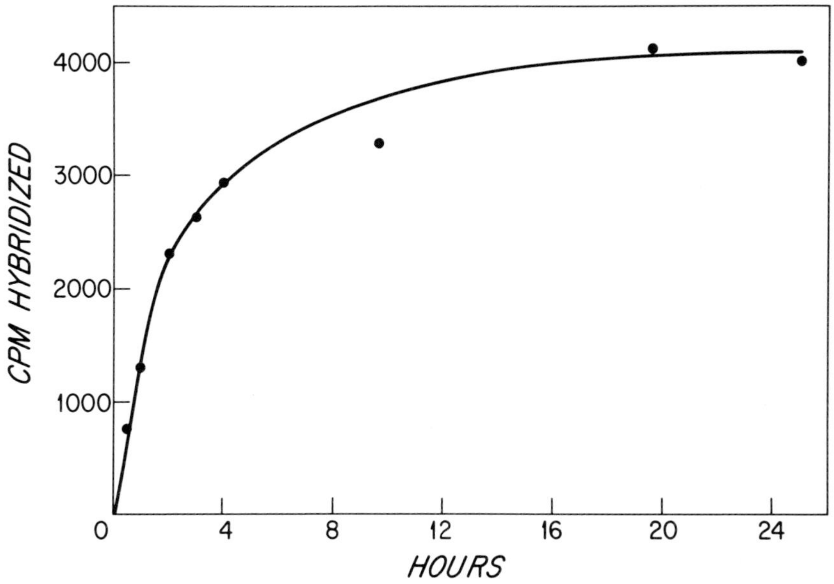

Fig. 1. Hybridization of fibroin mRNA with plasmid pBF39

DNA. Forty nanograms of pBF39 DNA, cleaved at a single site with restriction endonuclease Hind III, was bound to each filter. The hybridization solution consisted of 10 ml./filter of 2 nanograms/ml. fibroin mRNA-^{125}I.

The data, analyzed by the double reciprocal plot method (15, 16), extrapolate to saturation of pBF39 DNA with 1100 to 1300 nucleotides of mRNA per plasmid molecule. This corresponds to 4020 to 4750 cpm on the ordinate of Fig. 1. The time needed for hybridization to reach half of the saturation value varied inversely with RNA concentration over the range tested (2-20 ng. mRNA per ml.), as expected. Consequently it is possible to calculate the annealing reaction's $c_R t_{1/2}$ (the product of the RNA concentration in solution, in moles nucleotide residues per liter, and the time needed to attain half-saturation, in sec.). The $c_R t_{1/2}$ is $1.7 \pm 0.3 \times 10^{-4}$ M sec.

The kinetic complexity of the RNA annealing with pBF39 DNA is obtained by comparing its $c_R t_{1/2}$ to that of standard annealing reactions using RNAs of known complexity. The standards used were *Bacillus subtilis* and *Bombyx mori* ribosomal RNAs, annealing in RNA excess at the optimal temperature with the DNAs of their respective species. Their respective $c_R t_{1/2}$ values were found to be 1.2×10^{-2} M sec and 1.6×10^{-2} M sec. The analytic complexity (sum of molecular weights) of *B. subtilis* 23s and 16s rRNAs is 1.6×10^6 daltons, and the kinetic complexity of these RNAs is essentially the same (16).

We assume that the sum of the molecular weights of *Bombyx* 28s and 18s rRNAs is 2.15×10^6 (17), and that this is the kinetic complexity of a mixture of these rRNAs. $c_R t_{1/2} = N \cdot b$, where N is the kinetic complexity of the RNA and b is a constant (15, 16). We find $b = 7.5 \times 10^{-9}$ M sec/dalton. A published value for b is 8.4×10^{-9} M sec/dalton, obtained with ϕX174 DNA and complementary RNA, corrected to the molecular weight of ϕX174 from sequence analysis (16, 18). Using $b = 7.5 \times 10^{-9}$, the $c_R t_{1/2}$ of fibroin mRNA indicates that its kinetic complexity is 65 ± 15 nucleotides.

The plasmid pBF39 contains a representative sample of the DNA sequences of the fibroin gene. This is shown by the stability of fibroin mRNA-pBF39 hybrids. They melt at the same temperature as hybrids between fibroin mRNA and the fibroin gene in *Bombyx* DNA. The melting temperature is 59° in 70% formamide, 0.3 M NaCl, 0.06 M $NaPO_4$, pH6, 4 mM EDTA.

Large tandem duplications of the bacterial chromosome are unstable in wild type *E. coli* because they can be deleted by recombination events. They are stable in E. coli rec A^- strains (19, 20). Since pBF39 contains many tandem repeats of a DNA sequence, we examined its stability during growth in E. coli. To date only a rec A^- strain has been used in studying the stability of pBF39.

E. coli HB101 (pBF39) was spread on agar plates to obtain single colonies six times in succession, with growth in liquid culture intervening in some cases. Plasmid DNA was purified from cultures derived from single colonies at various stages extending to about 250 generations of growth after introduction of the plasmid DNA into the bacteria. Twenty preparations of DNA from single colonies were made. These were examined by digestion with a mixture of Hinc II and Hind III restriction endonucleases, which cleave pMB9 DNA at three sites and do not cleave the inserted fibroin sequence (1). The insertion in pBF39 is at the EcoRI site of pMB9, in the largest Hin II + III fragment of pMB9. Consequently the length of inserted sequence can be estimated by agarose gel electrophoresis of the Hin II + III digest to determine the molecular weight of the largest pBF39 DNA fragment (1). Eighteen of twenty pBF39 DNA preparations yielded a large fragment which appeared homogeneous in molecular weight, with an insertion of about 1380 base pairs. The other two preparations showed both an unchanged pBF39 large fragment and a fragment 700 base pairs smaller, presumably resulting from deletion. These fragments were extracted from the agarose gel by dissolving the gel slices in saturated KI solution and centrifuging to equilibrium in a KI gradient. Subsequent hybridization of these DNA fragments with fibroin mRNA in excess showed that they bound about one third as much mRNA as an equimolar quantity of the corresponding fragment from normal pBF39 DNA. This result suggests that the 700 base pairs of DNA deleted from pBF39 was fibroin DNA, but that some fibroin sequences were not deleted. Presumably about 400 base pairs remained.

The plasmid pBF39 and others derived from reverse transcripts of fibroin mRNA are useful for studies on the fibroin gene and RNA transcribed from it. However, they cannot provide information on the DNA sequences adjacent to the fibroin gene, which must contain DNA sequences recognized by other macromolecules to initiate transcription and to regulate the initiation of transcription. These can only be obtained from the DNA of the silkworm. A recombinant plasmid containing the DNA near the fibroin gene would provide quantities of DNA sufficient for biochemical analysis. It should be possible to find such a plasmid by screening a large number of clones containing Bombyx DNA for hybridization to fibroin mRNA (21). However, about 70,000 clones would have to be examined to provide a 90% probability that one containing the fibroin gene, a unique or single-copy gene (8, 12), would be found (22). We have attempted to reduce the labor involved by partial purification of the gene with neighboring DNA before formation of recombinant plasmids.

Since fibroin mRNA hybridizes primarily with the fibroin gene (8), it can serve as the basis for purification of the gene from other Bombyx DNA segments. Joining the messenger RNA to an insoluble matrix permits fast, simple purification of DNA hybridizing to it. We chose finely divided cellulose because DNA linked to it had been shown to be available for annealing (9). The cellulose particles are too small to be packed into a usable column, but they sediment rapidly when centrifuged (see EXPERIMENTAL).

The fibroin mRNA-cellulose anneals rapidly and specifically with the DNA of pBF39, the fibroin plasmid described above (Fig. 2).

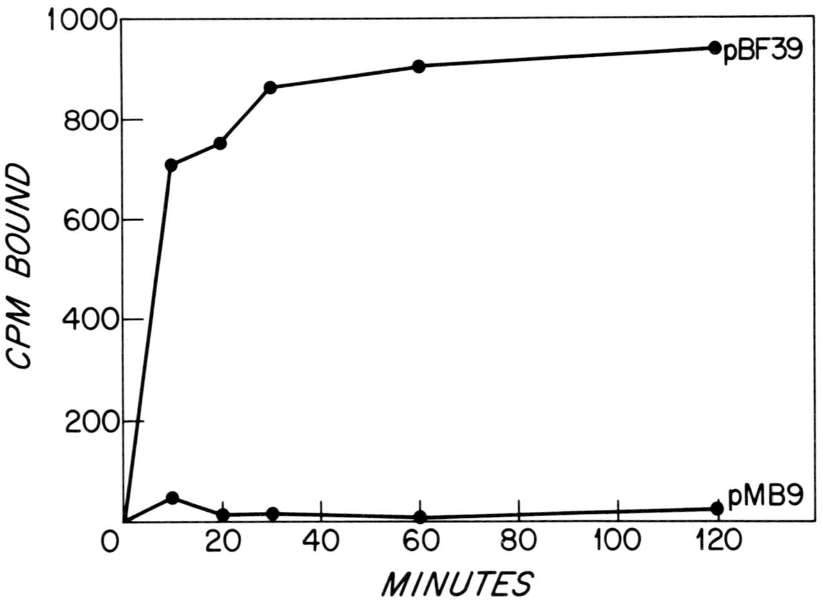

Fig. 2. Annealing of plasmid DNAs with fibroin mRNA-cellulose. Plasmid DNA-^{3}H cleaved at a single site with Hin$_{d}$III restriction endonuclease was denatured in alkali and neutralized. DNA at a concentration of 15 nanograms/ml. was annealed at 53° with mRNA-cellulose (0.15 μg/ml. of mRNA, 110 μg/ml. of cellulose) in the presence of 0.31 μg/ml. poly dT.

The excess of poly dT was included to prevent annealing between the poly A segment of fibroin mRNA and the poly dT linker segments of pBF39 DNA, synthesized by terminal transferase. The melting temperature of hybrids between pBF39 and mRNA-cellulose shows that the major, repeating sequence of fibroin mRNA annealed with the complementary sequence of pBF39. Melting was done in 70% formamide, 0.3 M NaCl, 0.06 M sodium phosphate, pH 6.0, 4 mM EDTA, by heating to a given temperature for 5 min., centrifuging, and counting the radioactive DNA in the supernate. The melting temperature is 62°, slightly higher than the melting temperature of fibroin mRNA-Bombyx DNA hybrids, which is 59° under these conditions. The high Tm is indicative of fibroin sequences, which are 60% G+C. The melting temperature of rA:dT is much lower.

The mRNA-cellulose anneals very rapidly with pBF39 DNA. The half-time for the reaction of Fig. 2 was about 5 min. Consequently the $c_R t_{1/2}$ was 2.6×10^{-4} M sec., comparable to the $c_R t_{1/2}$ for mRNA in solution annealing with pBF39 DNA on filters.

The rate of annealing of mRNA-cellulose with the fibroin gene of Bombyx DNA was also measured by a less direct experiment. Total Bombyx DNA was denatured (its average single-stranded molecular weight was 9000 nucleotides) and incubated with mRNA-cellulose. At intervals, the cellulose was sedimented, then washed, and the DNA eluted from it and bound to a nitrocellulose filter. DNA in the supernates was also bound to filters, and subsequently all were hybridized with 2 nanograms/ml. fibroin mRNA-^{125}I to determine the amount of fibroin gene in each fraction. Eighty per cent of the DNA capable of hybridizing fibroin mRNA was bound to mRNA-cellulose at the longer incubation times. The half-time for attainment of this plateau was about six minutes of annealing. The $c_R t_{1/2}$ was about 3.2×10^{-4} M sec.

Digestion of about 200 nucleotides from the ends of Bombyx DNA molecules permits many of those containing fibroin gene sequences to anneal to fibroin mRNA-cellulose. We have used exonuclease III of *E. coli* to remove residues from DNA 3′-termini, or phage lambda exonuclease to digest 5′-termini. Subsequent hybridization with mRNA-cellulose permits greater than one-thousand-fold purification of fibroin gene sequences from total Bombyx DNA. Our present efforts are directed toward incorporating this purified DNA into recombinant plasmids.

DISCUSSION

We have shown that the inserted DNA in plasmid pBF39 consists largely of the major repeated sequence of fibroin mRNA, presumably encoding -gly-ala-gly-ala-gly-ser- (8, 13). The kinetic complexity we found for the major sequence of fibroin mRNA, 65 ± 15 nucleotides, is consistent with an RNA sequence encoding a repeating hexapeptide sequence for two reasons. First, the RNA should repeat every 18 nucleotides, but the repetition is only approximate, not precise. Eight nucleotides of the eighteen may vary while the RNA still encodes -gly-ala-gly-ala-gly-ser-, because of the degeneracy of the genetic code. This variation causes mismatching between the complementary strands of different repeats (23). Mismatching decreases the rate of nucleic acid reassociation (24), thereby increasing the kinetic complexity from 18 nucleotides to perhaps 25 nucleotides in this case. Secondly, annealing rate may not vary inversely with sequence complexity when the sequence repeat length is less than the nucleic acid strand length used for rate measurements (25). In our measurements the RNA strand length was roughly 500 nucleotides, estimated from sedimentation rate at neutral pH. For these reasons, the kinetic complexities of satellite DNAs are greater than their repeat lengths from sequence analysis. For example, the kinetic complexities of mouse satellite and of guinea pig α-satellite are about 300 base pairs and several hundred thousand base pairs, respectively, while their sequence repeat lengths are about 18 and 6-9 base pairs (25-29). Kinetic complexity, 65 bases, is much closer to presumed sequence repeat length, 18 bases, for fibroin mRNA than for the satellite DNAs cited.

Stability of inserted sequences is very important for the use of bacterial cloning to study eukaryotic DNA segments. We are encouraged in this regard by the stability of pBF39, since few DNA sequences are more repetitious than the fibroin gene. However, it was recently reported that plasmids containing <u>Drosophila melanogaster</u> satellite DNA are unstable during growth in <u>E. coli</u> (30). The reason for the difference between that and our findings is not clear. Present results are consistent with the working hypothesis that any DNA sequence can be replicated as part of a recombinant plasmid. Even highly repetitious DNA sequences can be cloned in this way, though part of the DNA segment may be lost. Consequently supporting evidence for plasmid stability is important in studies on DNA sequence arrangement.

ACKNOWLEDGEMENTS

This work was supported in part by National Institutes of Health Grant GM22383. J. M. W. acknowledges the support of a U.S. Public Health Service Predoctoral Traineeship. A. E. was supported by an Intermediate Fellowship of the Harvard Society of Fellows.

REFERENCES

(1) J. F. Morrow, J. M. Wozney and A. Efstratiadis, in: Recombinant Molecules: Impact on Science and Society, ed. R. F. Beers, Jr. and E. G. Bassett (Raven Press, New York, 1977).

(2) A. Efstratiadis, T. Maniatis, F. C. Kafatos, A. Jeffrey and J. N. Vournakis, Cell, 4 (1975) 367.

(3) A. Efstratiadis, F. C. Kafatos, A. M. Maxam and T. Maniatis, Cell, 7 (1976) 279.

(4) T. Maniatis, S. G. Kee, A. Efstratiadis and F. C. Kafatos, Cell, 8 (1976) 163.

(5) R. A. Greene, M. Morgan, A. J. Shatkin and L. P. Gage, J. Biol. Chem., 250 (1975) 5114.

(6) R. L. Rodriguez, F. Bolivar, H. M. Goodman, H. W. Boyer and M. Betlach, in: Molecular Mechanisms in the Control of Gene Expression, eds. D. P. Nierlich, W. J. Rutter and C. F. Fox (Academic Press, New York, 1976) p. 471.

(7) W. Prensky, D. M. Steffensen and W. L. Hughes, Proc. Nat. Acad. Sci. USA, 70 (1973) 1860.

(8) Y. Suzuki, L. P. Gage and D. D. Brown, J. Mol. Biol., 70 (1972) 637.

(9) B. E. Noyes and G. R. Stark, Cell, 5 (1975) 301.

(10) M. V. Berridge and A. I. Aronson, Anal. Biochem., 53 (1973) 603.

(11) C. Aaij and P. Borst, Biochim. Biophys. Acta, 269 (1972) 192.

(12) P. Lizardi and D. D. Brown, Cell, 4 (1975) 207.

(13) F. Lucas and K. M. Rudall, in: Comprehensive Biochemistry, Vol. 26B, eds. M. Florkin and E. H. Stotz (Elsevier, New York, 1968) p. 475.

(14) Y. Suzuki and D. D. Brown, J. Mol. Biol., 63 (1972) 409.

(15) J. O. Bishop, Biochem. J., 113 (1969) 805.

(16) M. L. Birnstiel, B. H. Sells and I. F. Purdom, J. Mol. Biol., 63 (1972) 21.

(17) U. E. Loening, J. Mol. Biol., 38 (1968) 355.

(18) F. Sanger, G. M. Air, B. G. Barrell, N. L. Brown, A. R. Coulson, J. C. Fiddes, C. V. Hutchison, III, P. M. Slocombe and M. Smith, Nature, 265 (1977) 687.

(19) C. W. Hill, J. Foulds, L. Soll and P. Berg, J. Mol. Biol., 39 (1969) 563.

(20) W. R. Folk and P. Berg, J. Mol. Biol., 58 (1971) 595.

(21) M. Grunstein and D. S. Hogness, Proc. Nat. Acad. Sci. USA, 72 (1975) 3961.

(22) L. Clarke and J. Carbon, Cell, 9 (1976) 91.

(23) L. P. Gage and R. F. Manning, J. Mol. Biol., 101 (1976) 327.

(24) T. I. Bonner, D. J. Brenner, B. R. Neufeld and R. J. Britten, J. Mol. Biol., 81 (1973) 123.

(25) J. R. Hutton and J. G. Wetmur, Biochem. Biophys. Res. Comm., 52 (1973) 1148.

(26) M. Waring and R. J. Britten, Science, 154 (1966) 791.

(27) E. M. Southern, Nature, 227 (1970) 794.

(28) W. D. Sutton and M. McCallum, Nature, 232 (1971) 83.

(29) P. A. Biro, A. Carr-Brown, E. M. Southern and P. M. B. Walker, J. Mol. Biol., 94 (1975) 71.

(30) D. Brutlag, K. Fry, T. Nelson and P. Hung, Cell, 10 (1977) 509.

DISCUSSION

P. BERG: What estimate can you give as to the probability of producing a deletion in an inserted segment per generation?

J.F. MORROW: The majority of the DNA was taken through six streakings and a number of clones were picked after 250 generations. The majority of the DNA was unaltered as far as we could see.

P.L. GAGE: Have you examined the stability of the insert in a rec^+ background?

J.F. MORROW: We haven't looked at that yet. We are certainly very interested in doing it.

K. SAKAGUCHI: I would like to know whether the complementary DNA is homologous to the messenger RNA at every nucleotide residue? The product should be checked by S1 nuclease for hydrolysis at mismatched bases.

J.F. MORROW: We have not checked that but the data Dr. Maniatis will present in the next talk will speak more directly to this point.

AN APPROACH TO THE STUDY OF DEVELOPMENTALLY REGULATED GENES.

T. Maniatis[*†], G.K. Sim[*†], F.C. Kafatos[†],
L. Villa Komaroff[*†] and A. Efstratiadis[†].

[*]Cold Spring Harbor Laboratory,
Cold Spring Harbor, N.Y. 11724

[†]Harvard University Biological Laboratories
Cambridge, Massachusetts 02138

Abstract: A method has been developed for the *in vitro* synthesis of nearly complete double stranded DNA copies of polyA containing eukaryotic mRNAs. Amplification of *in vitro* synthesized DNA using molecular cloning techniques has provided large amounts of homogeneously pure eukaryotic gene sequences which can be studied directly or employed as hybridization probes for the identification and isolation of DNA sequences contiguous to structural genes in chromosomal DNA. The fidelity of *in vitro* synthesis and the stability of synthetic DNA in bacteria is demonstrated by the fact that the entire nucleotide sequence of a cloned DNA copy of rabbit β-globin mRNA is in complete agreement with available β-globin protein and mRNA sequence data. We are using the method of cDNA cloning to study the structure and chromosomal organization of a highly complex set of developmentally-regulated genes in the Silk Moth *Antharaea polyphemus*.

Introduction

The intractability of eukaryotic genomes to classical genetic analysis has provided the impetus for the development of alternative strategies for studying genetic regulation. One of the most fruitful approaches has been the isolation and characterization of specific eukaryotic genes (see 1 for review). Until recently, genes have been isolated on the basis of their relatively high G+C content and tandemly repetitious organization. The most thoroughly studied examples of this are the ribosomal genes of *Xenopus laevis* (2,3) and the histone genes of sea urchins (4,5). More recently, molecular cloning techniques have made it possible to isolate and characterize eukaryotic genes irrespective of their base composition. A wealth of information regarding the sequence organization of ribosomal and histone gene clusters has been derived from cloned eukaryotic DNA (6, 7,8,9).

In principle, any eukaryotic gene can be isolated using molecular cloning techniques, but in practice, the isolation of single copy structural genes from mammalian cells is a formidable technical problem. For example, the rabbit β-globin gene of approximately 600 base pairs represents only 2×10^{-5}% of the haploid genome. Thus, only 0.2 ug of the gene is present in one gram of rabbit DNA. If rabbit chromosomal DNA is sheared into fragments of approximately 15 kilobases, joined to a bacteriophage lambda or plasmid DNA vector and cloned, it would be necessary to screen nearly one million clones containing hybrid DNA to be certain of identifying the plaque or colony carrying the β-globin gene (see 10 for discussion of this point). *In situ* hybridization procedures have been developed for screening large numbers of plaques containing chimeric bacteriophage lambda DNA (11) or bacterial colonies containing hybrid plasmid DNA (12). However, large amounts of highly purified hybridization probe is required and, for most mRNAs, this is difficult or impossible to obtain.

We have developed a procedure for cloning and amplifying eukaryotic mRNA sequences for the purposes of preparing large amounts of homogeneous gene sequences which can be used for the isolation and characterization of DNA sequences in chromosomal DNA (13-15) and for studying the primary structure and evolution of mRNA. The procedure involves reverse transcription of the mRNA under conditions favoring the production of cDNA copies equal to the template length (13), synthesis of a second DNA strand from a hairpin loop at the 3'-end of the cDNA using *E. coli* polymerase I; cleavage of the loop joining the two strands with SI nuclease (14); insertion of the synthetic duplex DNA into a bacterial plasmid by the dA-dT joining procedure; and transformation of bacteria with the hybrid DNA molecules (15). Other laboratories have developed similar or somewhat different approaches (16-18).

We have monitored the accuracy and completeness of the *in vitro* synthesis of double stranded cDNA at every step of the procedure using nucleic acid hybridization techniques, restriction endonuclease cleavage analysis, and high resolution native and denaturing polyacrylamide gels (13-15). Figure 1 shows an example of the characterization of one plasmid, PβG-1, which carries a nearly complete copy of β-globin mRNA. The position of the globin insertion with respect to a single *HindIII* cleavage site in the hybrid plasmid DNA is revealed by hybridization to globin mRNA under conditions which favor the formation of an R-loop (19,20). In the appropriate conditions, globin mRNA hybridizes to the double stranded plasmid DNA and displaces the message synonomous strand. A detailed physical map of the globin insertion in PβG-1 is shown in Figure 1B. This restriction map is identical to that of the synthetic β-globin DNA and can be aligned with the mRNA sequence predicted from the amino acid sequence of β-globin (15). Thus, at the level of restriction endonuclease cleavage sites, the cloned globin DNA represents a faithful copy of the mRNA and no sequence rearrangements occurred as a result of the cloning procedures or subsequent propagation of the hybrid plasmid DNA in bacteria.

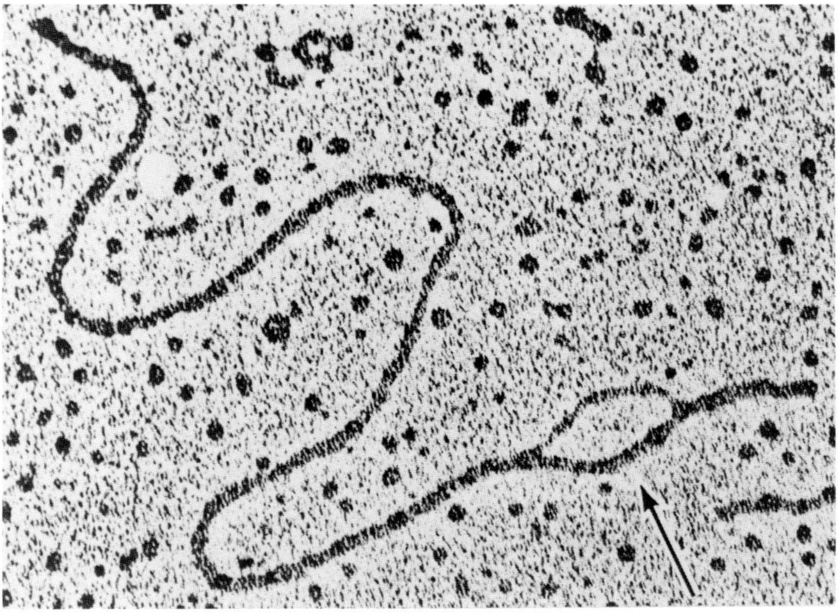

Figure 1A Electron microscopic visualization of the β-globin insertion in PβG-1 DNA by R-loop formation.

PβG-1 DNA was linearized by digestion with the restriction enzyme HindIII which cleaves at a single site near the β-globin insertion. Following phenol extraction and ethanol precipitation, the DNA was resuspended and mixed with purified globin mRNA. The mixture (4.5 ug/ml DNA, 4.7 ug/ml RNA) was incubated for 12 hrs in 70% formamide, 0.4M NaCl, 0.1M PIPES pH 7.8 at 52° and spread in 70% formamide. The arrow indicates the position of the globin mRNA-DNA duplex. Magnification: 160,000 X. The picture was taken by L. Chow and T. Broker of the Cold Spring Harbor Laboratory.

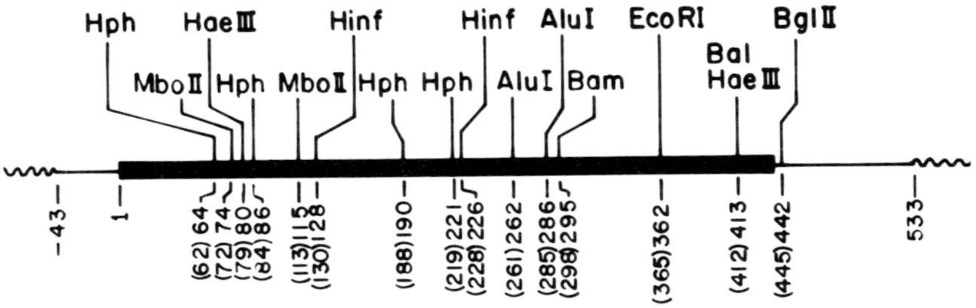

Figure 1B Restriction endonuclease cleavage map of the β-globin insertion in PβG-1.

The coding sequence of β-globin DNA (thick horizontal line), the untranslated sequences (thin horizontal line) and the poly (dA)-poly (dT) bridges which join the globin DNA to the plasmid (wavy lines) are indicated. Each restriction site is identified by numbers indicating the 5' terminal nucleotide generated by cleavage of the message strand (plain numbers) of the anti-message strand (numbers in parentheses). As discussed elsewhere (30), all nucleotides are numbered by reference to the first nucleotide following the initiation codon of β-globin mRNA.

DNA Sequence Analysis

As mentioned above, one of the potential uses of gene amplification procedures is to determine the primary structure of eukaryotic mRNAs. Because of difficulties in obtaining *in vivo* labeled mRNAs at specific activities sufficient for direct nucleotide sequence analysis, a number of indirect sequencing methods have been developed (21,22,23,24). These methods involve the analysis of *in vitro* labeled mRNA, cDNA or cRNA using established RNA or DNA sequencing methods. Unfortunately, large amounts of highly purified mRNA is required, and in most cases, is not available. In addition, the indirect methods have so far provided relatively short stretches of sequence adjacent to the 3'-terminus of the mRNA or a catalog of T1 oligonucleotides which cannot be ordered without reference to an amino acid sequence. By contrast, cloned DNA copies of mRNA can be prepared in unlimited amounts. The amplified DNA can then be cleaved into many short segments with restriction enzymes and sequenced using the rapid DNA sequencing method of Maxam and Gilbert (25) or Sanger and Coulson (26). However, before this approach to the study of mRNA structure can be used with confidence, agreement between sequences derived from cloned cDNA and independently derived mRNA sequence data must be demonstrated.

The β-globin plasmid (PβG-1) is well suited for such a comparative analysis. The entire amino acid sequence of - globin is known (27), and extensive partial mRNA sequencing data is available (21,22,28,29). Using the detailed restriction map of PβG-1 (Figure 1B) as a guide and the Maxam-Gilbert sequencing method, we have derived the entire 576 base pair sequence of the β-globin insertion of PβG-1. The details of this work have been presented elsewhere (30). Only the essential features will be discussed here.

The Maxam-Gilbert procedure depends on the availability of DNA fragments, each uniquely labeled with ^{32}P at a single 5'-end. Defined double stranded DNA fragments can be generated by restriction endonuclease digestion and both 5'-ends labeled *in vitro* using γ-^{32}P-ATP and T4 polynucleotide kinase. The two labeled ends can be separated by secondary cleavage with another restriction enzyme followed by purification of the singly labeled DNA fragments on polyacrylamide gels. Alternatively, the denatured DNA can be separated by polyacrylamide gel electrophoresis (25). The fragments bearing a single terminal label are subjected to four sets of chemical reactions. These are base specific modifications leading to base removal and ultimately strand scission. With the appropriate conditions, each molecule is cleaved at a single base specific, but random position in the chain. Thus, four different sets of labeled products are generated (one from each base) all beginning at the uniquely labeled 5'-end but terminating at various positions, depending on the base eliminated. The samples are then subjected to electrophoresis in parallel under denaturing conditions in polyacrylamide gels capable of resolving products differing in chain length by one nucleotide. The DNA sequence can then be read from the autoradiogram.

An example of the sequencing data is shown in Figure 2. The sequence shown extends from the RI site at position 365 to the *AluI* site at position 286 in the anti-message strand of the globin DNA insertion in PβG-1. The labeled DNA fragment was prepared by digesting PβG-1 DNA with RI, end-labeling, digesting with *AluI* and purifying the 79 base pair fragment by polyacrylamide gel electrophoresis. Aliquots of this DNA were subjected to the four types of base specific chemical reactions and the cleavage products analyzed on a denaturing polyacrylamide gel. Resolution of different regions of the sequence was maximized by layering aliquots of the four samples on adjacent gel slots at different times after the beginning of electrophoresis (25). In this way, the first 30-40 nucleotides can be derived from the samples electrophoresed for the shorter time period while nucleotide 25 through 77 can be read from the samples which are run longer. The first three nucleotides which are not shown on the autoradiogram can be deduced from the RI recognition sequence and the last two nucleotides in the fragment can be deduced from the recognition sequence of AluI. The remainder of the globin sequence of Pβ G-1 was derived in a similar manner and is described in detail elsewhere (30).

Once the entire β-globin sequence of Pβ G-1 was derived, it was possible to establish the validity of this approach for to the sequence analysis of eukaryotic mRNAs by comparing the plasmid DNA sequence with indepently-derived mRNA sequencing data and the mRNA sequence expected from the amino acid sequence of β-globin. This comparison is outlined in Figure 3 and can be summarized as follows:

1. Comparison with published mRNA sequencing data.

Proudfoot and Brownlee (21) have determined the sequence of 164 nucleotides near the 3'-end of β-globin mRNA. These sequences extend from position 321 to 408 and 459 to 533, as shown in Figure 3, and are identical to the corresponding region of the sequence we derived from PβG-1. By direct RNA sequencing methods, Lockhard and RajBhandary (24) have determined the 5'-terminal sequence of 26 nucleotides immediately adjacent to the 7Methyl guanine residue of β-globin mRNA. As shown in Figure 3, PβG-1 is missing only 13 nucleotides at the extreme 5'-end of this sequence and is identical to the remainder. The PβG-1 sequence also agrees with a stretch of 19 nucleotides in position -15 to +5 in Figure 3 identified by Legon (28) as the ribosome binding site of β-globin mRNA. Finally, Salser and his collaborators (22) have published an extensive catalog of Tl oligonucleotides of RNA copied from α and β rabbit globin cDNA and have positioned some of these fragments within the coding sequence. Figure 3 shows these assignments plus additional Tl fragments which can now be matched unambiguously with segments of the β-globin mRNA sequence. The published catalog includes additional Tl fragments which cannot be matched and presumably correspond to α-globin mRNA.

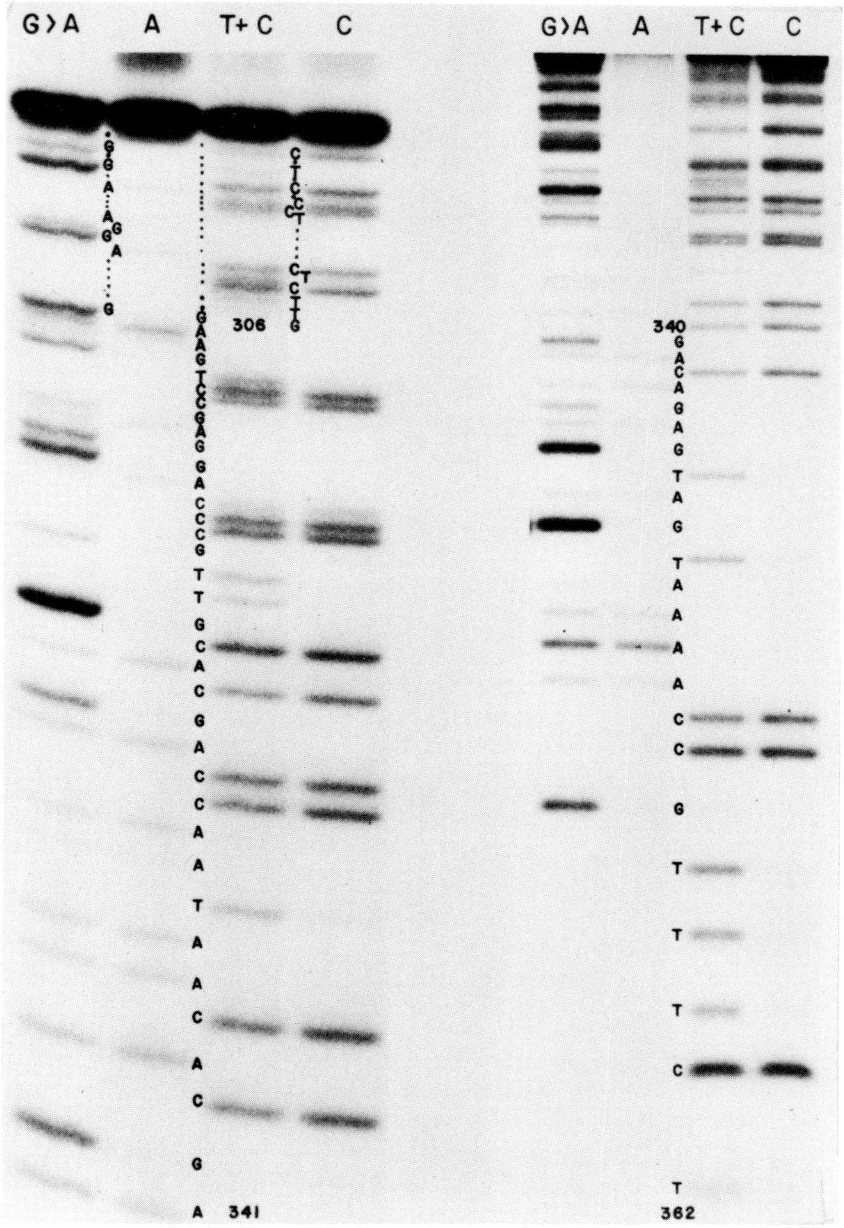

Figure 2 An example of the data used to derive the β-globin sequence in PβG-1.

A small (79 base pair) DNA fragment labeled at nucleotide 365 (see Figure 1A) was subjected to four sets of chemical reactions and the base-specific degradation products analysed by electrophoresis on a 20% polyacrylamide 7M urea gel. Identical sets of the four samples were run for different periods of time. The figure shows the autoradiogram from the shorter (right) and longer (left) electrophoretic runs together with the inferred sequence.

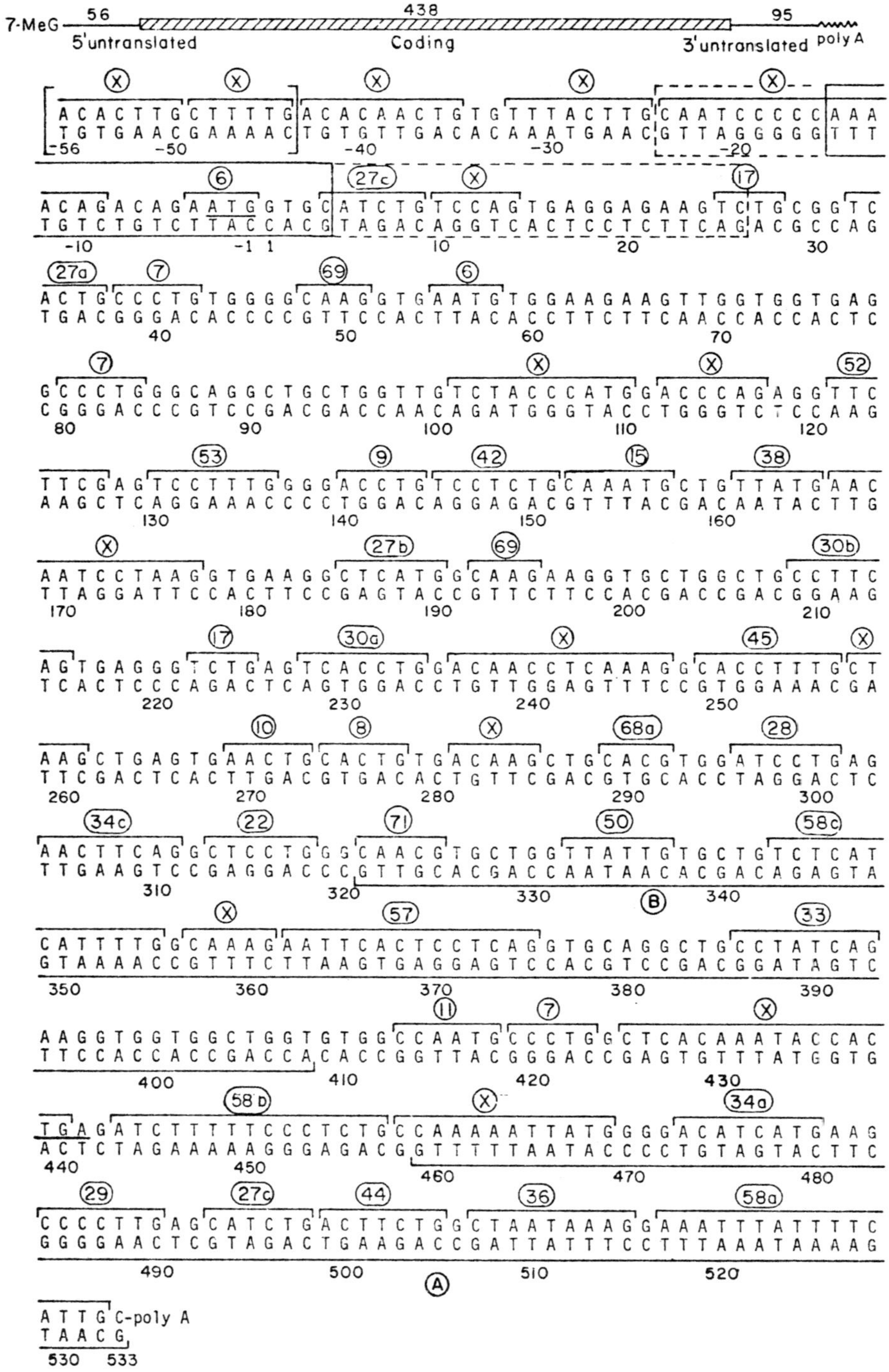
7-MeG 56 438 95
5'untranslated Coding 3'untranslated poly A
X X X X X
ACACTTGCTTTTGACACAACTGTGTTTACTTGCAATCCCCCAAA
TGTGAACGAAAACTGTGTTGACACAAATGAACGTTAGGGGGTTT
-56 -50 -40 -30 -20
6 27c X 17
ACAGACAGAATGGTGCATCTGTCCAGTGAGGAGAAGTCTGCGGTC
TGTCTGTCTTACCACGTAGACAGGTCACTCCTCTTCAGACGCCAG
-10 -1 1 10 20 30
27a 7 69 6
ACTGCCCTGTGGGGCAAGGTGAATGTGGAAGAAGTTGGTGGTGAG
TGACGGGACACCCCGTTCCACTTACACCTTCTTCAACCACCACTC
40 50 60 70
7 X X 52
GCCCTGGGCAGGCTGCTGGTTGTCTACCCATGGACCCAGAGGTTC
CGGGACCCGTCCGACGACCAACAGATGGGTACCTGGGTCTCCAAG
80 90 100 110 120
53 9 42 15 38
TTCGAGTCCTTTGGGGACCTGTCCTCTGCAAATGCTGTTATGAAC
AAGCTCAGGAAACCCCTGGACAGGAGACGTTTACGACAATACTTG
130 140 150 160
X 27b 69 30b
AATCCTAAGGTGAAGGCTCATGGCAAGAAGGTGCTGGCTGCCTTC
TTAGGATTCCACTTCCGAGTACCGTTCTTCCACGACCGACGGAAG
170 180 190 200 210
17 30a X 45 X
AGTGAGGGTCTGAGTCACCTGGACAACCTCAAAGGCACCTTTGCT
TCACTCCCAGACTCAGTGGACCTGTTGGAGTTTCCGTGGAAACGA
220 230 240 250
10 8 X 68a 28
AAGCTGAGTGAACTGCACTGTGACAAGCTGCACGTGGATCCTGAG
TTCGACTCACTTGACGTGACACTGTTCGACGTGCACCTAGGACTC
260 270 280 290 300
34c 22 71 50 58c
AACTTCAGGCTCCTGGGCAACGTGCTGGTTATTGTGCTGTCTCAT
TTGAAGTCCGAGGACCCGTTGCACGACCAATAACACGACAGAGTA
310 320 330 B 340
X 57 33
CATTTTGGCAAAGAATTCACTCCTCAGGTGCAGGCTGCCTATCAG
GTAAAACCGTTTCTTAAGTGAGGAGTCCACGTCCGACGGATAGTC
350 360 370 380 390
11 7 X
AAGGTGGTGGCTGGTGTGGCCAATGCCCTGGCTCACAAATACCAC
TTCCACCACCGACCACACCGGTTACGGGACCGAGTGTTTATGGTG
400 410 420 430
58b X 34a
TGAGATCTTTTTCCCTCTGCCAAAAATTATGGGGACATCATGAAG
ACTCTAGAAAAAGGGAGACGGTTTTTAATACCCCTGTAGTACTTC
440 450 460 470 480
29 27c 44 36 58a
CCCCTTGAGCATCTGACTTCTGGCTAATAAAGGAAATTTATTTTC
GGGGAACTCGTAGACTGAAGACCGATTATTTCCTTTAAATAAAAG
490 500 A 510 520
ATTGC-poly A
TAACG
530 533

Figure 3 Correlation between the globin sequence in Pβ G-1 DNA and published β-globin mRNA sequencing data.

The distribution of β-globin mRNA sequences in the coding and non-coding regions (including initiator and terminator codon) are shown schematically at the top of the figure. The double strand DNA sequence corresponding to the entire mRNA is shown (top strand synonymous with the mRNA). The base pairs are numbered from -56 to 533, with nucleotide 1 being the first nucleotide following the initiator AUG. The region in brackets from position -56 to -44 is the sequence derived directly from the mRNA by Lockard and RajBhandary but missing in Pβ G-1. Nucleotides in positions -44 to -31 are identical to the sequence derived by Lockard and RajBhandary (29, personal communication). The region enclosed with solid lines (-15 to +4) is identical to the ribosome binding sequence reported by Legon (28). The region from position 5 to 24 (enclosed with dashed lines) is thought to approximately define the 3' limit of the 80S ribosome binding sequence. The region enclosed with a dashed line extending to position -24 is included in the 40S ribosome binding site. The underlined sequences ATG and TGA, at positions -3 to -1 and 439 to 441 identify the initiator and terminator codons, respectively. All of the oligonucleotides of 4 nucleotides or greater which would be expected to arise from a ribonuclease T1 digestion of β-globin mRNA are indicated by brackets over the message strand. Each of the predicted T1 oligonucleotides is labeled with either a number corresponding to the published T1 catalogue of alpha and beta globin mRNA (22) or by an X which indicates the absence of the oligonucleotide from the catalogue. Nucleotides 459 to 469 may possibly correspond to T1 fragment 55. The identification of nucleotides 443 to 457 with fragment 58b is tentative, since the latter is listed in the catalogue as AUC(Y_{7-8}) CUCUG. The two lines below the sequence (A and B position 320 to 408 and 459 to 533, respectively) indicate regions identical to those sequenced by Proudfoot (42, 43). Position 341 to 419 also agree with a partial sequence determined by Salser et al. (44).

2. Comparison with nearest-neighbor pyrimidine tract data.

We have carried out a complete nearest-neighbor pyrimidine tract analysis of in vitro labeled double stranded DNA copied from β-globin mRNA as a means of obtaining additional mRNA sequence information for comparison to the β-globin sequence derived from cloned DNA. To accomplish this, four different types of in vitro labeled double stranded globin DNA was prepared, each one labeled in only one strand with either α-^{32}P-dATP or α-^{32}P-dGTP. Highly purified globin mRNA was copied into double stranded DNA by performing in sequence reverse transcription and DNA polymerase I reactions, using a single ^{32}P-labeled deoxynucleotide in only one of the enzyme reactions. The DNA preparations were treated with S1 nuclease to cleave the loop connecting the strands (14) and the full length β-globin DNA purified from partial and α-globin DNA transcripts by polyacrylamide gel electrophoresis as shown in Figure 4. The band designated 580 was excised, the DNA eluted, digested with various restriction endonucleases and specific fragments from different regions of the sequence isolated. Depurination analysis was performed on the intact DNA (Figure 5) and on individual restriction fragments (Figure 6). Upon depurination and strand scission of the DNA, the phosphate of purine nucleotides is transferred to the 5' nearest-neighbor pyrimidine tract. Thus the availability of the four preparations of globin DNA made it possible to examine separately the sets of labeled pyrimidine tracts of the message and anti-message strands adjacent to either A or G. Figure 7 illustrates how the pyrimidine tract analysis was used to test the validity of the sequence derived from Figure 2. The dots above the sequence in Figure 7 indicate those nucleotides which, if changed, would result in a qualitative change in at least one of the four pyrimidine tract fingerprints of Figure 6. For example, if nucleotide 300 in the message strand was C, rather than T, a TC_3 would have appeared among the depurination products in panel A of Figure 6 and the T_2C_2 tract would be missing. A similar analysis of the other nucleotides in Figure 7 indicates that an incorrect identification of a purine or pyrimidine on the sequencing gel would be revealed in 54 out of 79 nucleotides by either a missing or an additional spot in one of the four pyrimidine tract fingerprints. Thus, for small DNA fragments, the depurination data provide a powerful means of sequence verification. As discussed elsewhere (30), the depurination data derived from DNA copies of the mRNA is in complete agreement with the globin sequence in PβG-1.

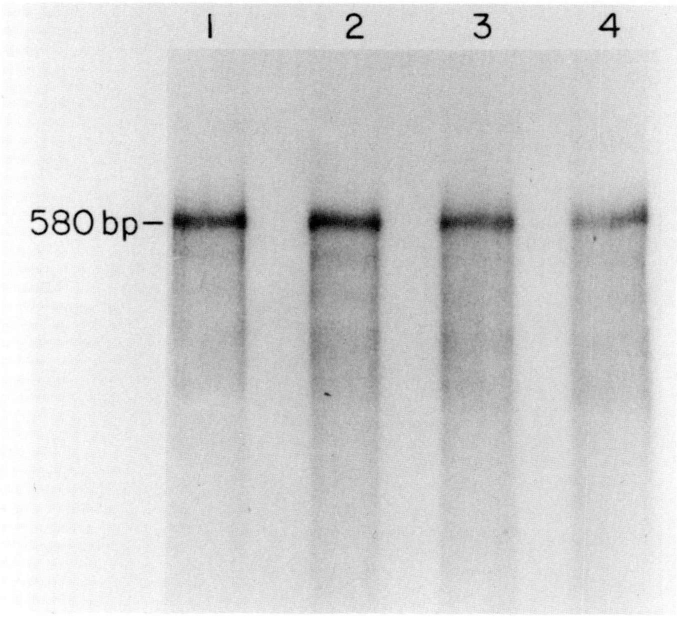

Figure 4. Autoradiogram of in vitro labeled double stranded globin cDNA fractionated on a 4% polyacrylamide gel.

Double stranded cDNA was synthesized in vitro as previously described (14) except that labeled precursor (α-^{32}P-dATP or α-^{32}P-dGTP) was included in either the first strand (anti-message) or second strand (message) synthetic reactions but not in both. Thus, each set of double stranded DNA molecules is labeled in only one strand. The synthetic reaction products were digested with S1 nuclease, ethanol precipitated and fractionated on a 4% polyacrylamide gel as described previously (15).

The bands labeled 580 bp represent complete copies of β-globin mRNA. These bands were excised from the gel and analysed as described in the text.

(1) Antimessage strand labeled with α-^{32}P-dATP.

(2) Antimessage strand labeled with α-^{32}P-dGTP.

(3) Message strand labeled with α-^{32}P-dGTP.

(4) Message strand labeled with α-^{32}P-dATP.

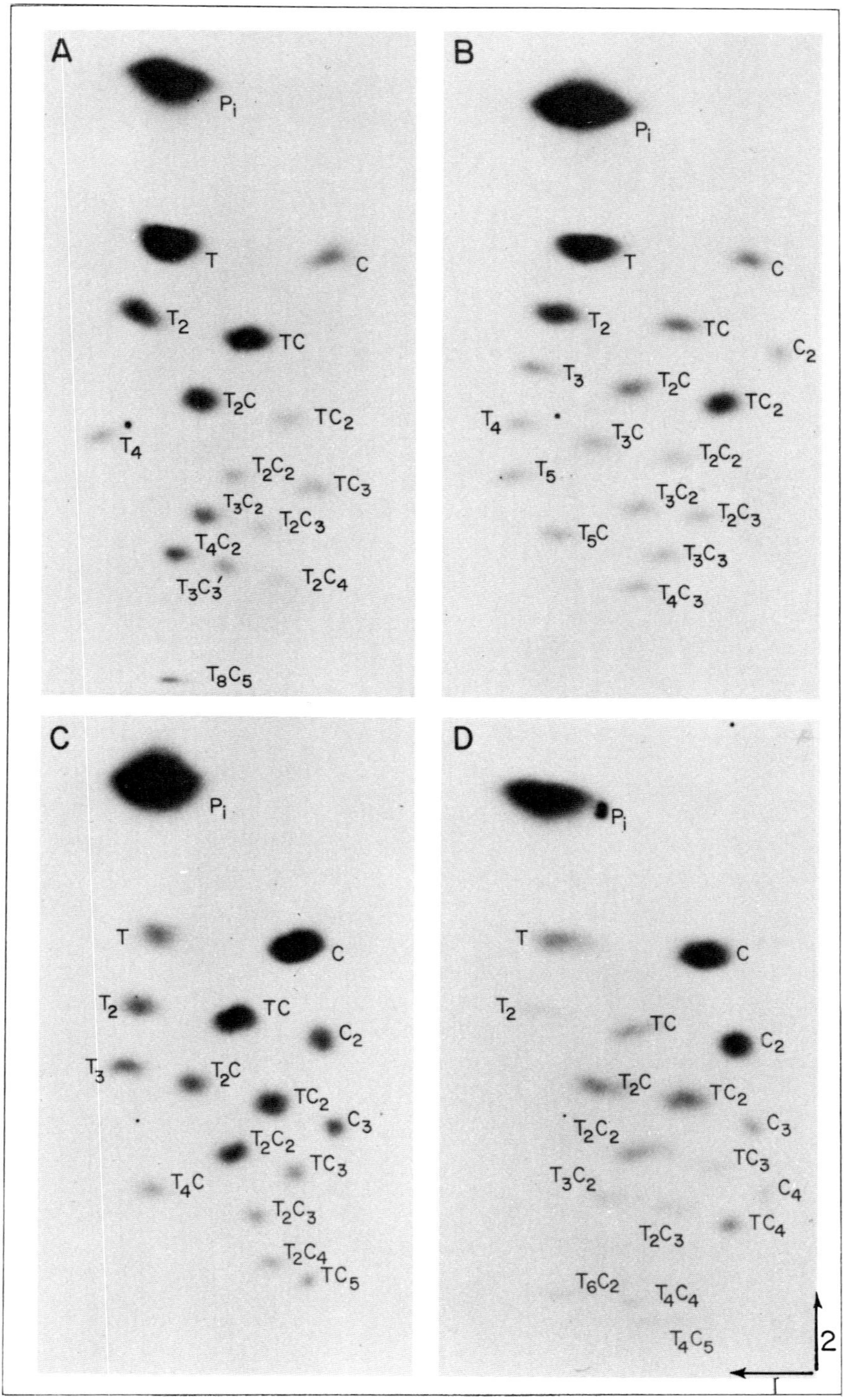
A
P_i
T
C
T_2
TC
T_2C
TC_2
T_4
T_2C_2
TC_3
T_3C_2
T_2C_3
T_4C_2
T_3C_3'
T_2C_4
T_8C_5
B
P_i
T
C
T_2
TC
C_2
T_3
T_2C
TC_2
T_4
T_3C
T_2C_2
T_5
T_3C_2
T_2C_3
T_5C
T_3C_3
T_4C_3
C
P_i
T
C
T_2
TC
C_2
T_3
T_2C
TC_2
C_3
T_2C_2
TC_3
T_4C
T_2C_3
T_2C_4
TC_5
D
P_i
T
C
T_2
TC
C_2
T_2C
TC_2
C_3
T_2C_2
TC_3
T_3C_2
C_4
TC_4
T_2C_3
T_6C_2
T_4C_4
T_4C_5
2
1

Fig. 5

Figure 5 Autoradiograms of pyrimidine tract fingerprints of *in vitro* synthesized globin DNA.

Double stranded globin DNA was labeled with either α-^{32}P-dATP or α-^{32}P-dGTP in the message or anti-message strand (see Figure 4) depurinated with formic acid and diphenylamine and the products fractionated by two-dimensional homochromatography (31) using the homomixture VI of Jay et al. (45). The composition of the pyrimidine tracts was determined by comparison to a complete set of tracts derived from bacteriophage DNA and fractionated in parallel. All tracts could be unambiguously identified except for T_8C_5 of sample A which is too large to be resolved by the homomixture VI. Arrows indicate the first (horizontal) and second (vertical) dimensions.

(A) Message strand labeled by α-^{32}P-dGTP.

(B) Antimessage strand labeled by α-^{32}P-dGTP.

(C) Message strand labeled by α-^{32}P-dATP.

(D) Antimessage strand labeled by α-^{32}P-dATP.

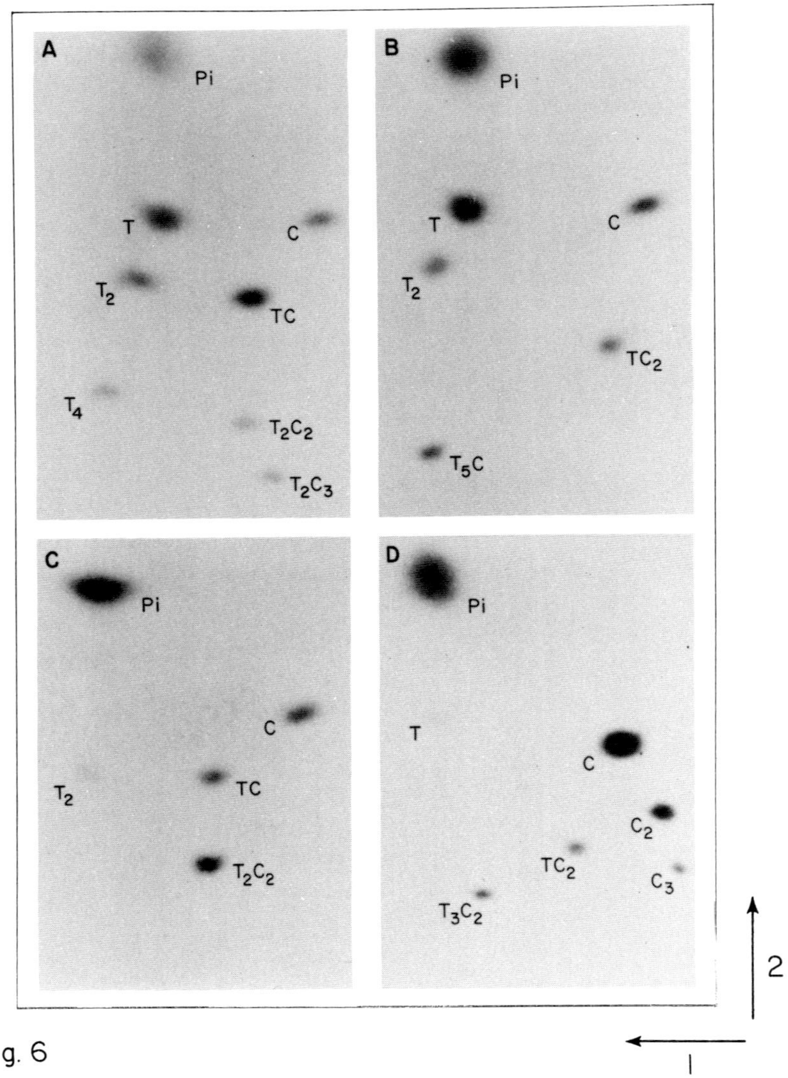
A
Pi
T
C
T2
TC
T4
T2C2
T2C3
B
Pi
T
C
T2
TC2
T5C
C
Pi
C
TC
T2
T2C2
D
Pi
T
C
C2
TC2
C3
T3C2
2
1

Fig. 6

Figure 6 Autoradiograms of pyrimidine tract fingerprints of a 79 bp globin DNA fragment.

The 580 bp globin DNA isolated from the polyacrylamide gel of Figure 6 was subjected to digestion by a number of restriction endonucleases including RI and Alu, fractionated on a 5% polyacrylamide gel and the individual digestion products recovered, depurinated and the pyridimine tracts fractionated by two-dimensional homochromatography. The example shown is the 79 base pair fragment of Figure 2.

(A) Message strand labeled with α-^{32}P-dGTP.

(B) Antimessage strand labeled with α-^{32}P-dGTP.

(C) Message strand labeled with α-^{32}P-dATP.

(D) Antimessage strand labeled with α-^{32}P-dATP.

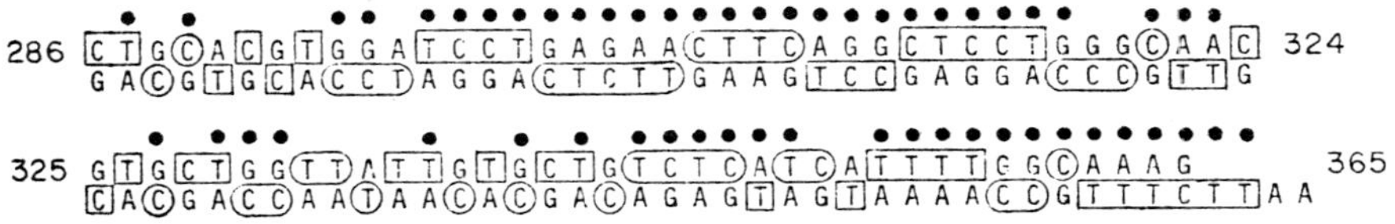

Figure 7 Example of sequence verification by pyrimidine tract analysis.

The double stranded DNA sequence of the 79 bp fragment of figures 2 and 6 is shown. Pyrimidine tracts which are nearest neighbor to guanine are shown in boxes while those which are nearest neighbor to adenine are circled. The dots above the sequence indicate those nucleotides which, if changed to the alternative purine or pyrimidine (A G or T C), would result in a qualitative difference in the pyrimidine tracts in one of the 4 panels of figure 6.

3. Comparison with the mRNA sequence predicted from the primary structure of β-globin.

The nucleotide sequence we have derived for PβG-1 is in perfect agreement with the mRNA sequence predicted from the primary structure of β-globin (30). Furthermore, the cloned DNA codes for ASN at amino acid positions 52 and 56, SER at position 76, and ILE at position 112. This is in agreement with the known structure of one of the common β-globin alleles, the alternative being 52 HIS, 56 SER, 76 ASN, and 112 VAL (32).

In conclusion, the β-globin sequence of PβG-1, which includes all but the initial 13 5'-nucleotides of the β-globin mRNA sequence, is in complete agreement with all known protein and mRNA sequencing data. 444 out of 589 nucleotides of β-globin mRNA can be verified on the basis of existing β-globin protein and unambiguous mRNA sequencing data. However, considering both the nearest neighbor pyrimidine tract data, which was derived from double stranded cDNA, and the published T1 oligonucleotide data, approximately

97% of the total β-globin sequence in PβG-1 can be verified with independently derived protein and nucleic acid sequence data. The absence of any conflicts provide strong evidence for the validity of using recombinant DNA for structural studies of eukaryotic genes. The interesting features of the sequence, which include the selective use of particular codons, and the probable existence of extensively base paired segments of the 5'-terminal region of the mRNA, are discussed in detail elsewhere (30).

The Moth Chorion as a system for studying the structure of developmentally regulated genes.

We have recently begun studying the structure of a highly complex set of developmentally regulated genes which code for the proteins forming the eggshell (chorion) in the silk moth *Antheraea polyphemus* (see 33 for review). In this animal, maturing eggs are found in 8 linear arrays (see Figure 8). Each array is called ovariole and each egg plus its accessory structures is called a follicle. The ovariole is essentially an assembly line for follicle maturation so that the spatial dimension of the ovariole corresponds to the temporal dimension of follicle development. Adjacent follicles are separated by approximately 4 hrs of developmental age difference and can be staged approximately by their position within the array and exactly by reference to the precise pattern of proteins they synthesize (34). The ultra-structure of the chorion is highly complex. On the basis of two-dimensional protein gel analysis, as many as 50 different chorion proteins can be identified (unpublished results). Most of these are low molecular weight (7000-20,000 daltons), acidic proteins which can be classified into four major size classes. The proteins are very rich in the nonpolar amino acids glycine (33%) and alanine (12%) and are also rich in cystine, 6% (35). Amino acid sequence data indicates that the proteins are highly homologous, suggesting that they have evolved by gene duplication followed by sequence divergence. On the basis of this data, at least one group of chorion proteins fulfill most of the criteria defining a multi-gene family. These are functionally and evolutionarily related genes which are physically linked and evolve rapidly in species-specific clusters (36).

Purified chorion mRNAs appear as two broad zones after electrophoresis on polyacrylamide gels (37). The mRNAs have been shown to program cell-free translation systems for the production of chorion proteins as assayed by chorion-specific antibody precipitation (38), and are synthesized and accumulate only during choriogenesis (39). 97% of chorion mRNA is found exclusively in polysomal and ribosomal fractions.

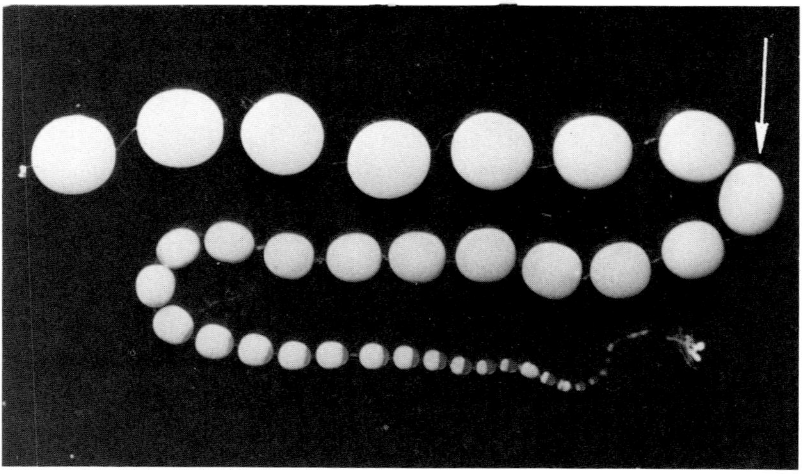

Figure 8. An ovariole dissected from a developing *Antheraea polyphemus* moth.

The maturing follicles are interconnected by a cord of cells and form a linear array, from bottom right to top left. The arrow indicates a follicle during terminal growth, just before the onset of choriogenesis. The diameter of the most mature follicle measures approximately 2.7mm.

Because of their similar molecular weights and variable amounts of polyA sequences at the 3'-ends, it has not been possible to resolve chorion mRNAs into single species by polyacrylamide gel electrophoresis. Specific removal of polyA by digestion with RNAse H in the presence of oligo dT convert each of the two broad zones of chorion mNRA into a series of sharper bands slightly lower in molecular weight but still does not provide the required resolution (40).

By using the procedures for gene cloning outlined for globin, it is possible to obtain individual chorion mRNA sequences in homogeneous form and in large amounts. By synthesizing double stranded cDNA from a heterogeneous population of chorion mRNA molecules, inserting these sequences into plasmids, and cloning them in bacteria, it is possible to obtain bacteria which carry chorion DNA sequences corresponding to only one of the mRNA species present in the mixture of mRNA.

Starting with purified chorion mRNA, isolated from the entire set of choriogenic follicles, we have synthesized double stranded cDNA and purified the major components by polyacrylamide gel electrophoresis. The DNA was eluted from the gel and inserted into the bacterial plasmid PML-21 which carries a kanamycin drug-resistance marker. The hybrid plasmid DNA molecules were then used to transform E. coli to kanamycin resistance. Colonies carrying plasmids with chorion mRNA sequences were identified by hybridization to ^{32}P-labeled chorion cDNA in situ (12). On the basis of the estimates of the number of proteins synthesized during choriogenesis, we anticipated that as many as 50 different chorion mRNA sequences are present in the collection of cloned chorion DNA sequences. In order to use these as hybridization probes for studying the organization of chorion genes, it is essential to compile a catalog of individual mRNA sequences. To accomplish this, we have used two approaches:

(1) Restriction endonuclease cleavage analysis. We have compared the restriction endonuclease digestion products of independently isolated plasmid DNA molecules by gel electrophoresis. We started by using restriction enzymes with recognition sequences of 5 or 6 base pairs which cleave the relatively small plasmid DNA molecules once or a few times. Thus, we could group hybrid plasmid DNA molecules into different classes on the basis of whether or not the chorion-DNA insertion of that plasmid was cleaved by a particular enzyme. In order to distinguish between individual chorion gene plasmids within these major groups, other restriction enzymes which recognize tetranucleotide sequences, and therefore cleave the plasmid DNA many times, were used in the analysis. By comparing the digestion patterns of plasmid DNAs within the major groups, it was possible to identify a large number of "different" plasmids. One difficulty with using enzymes which cleave the DNA into many fragments is that the smaller chorion DNA fragments are sometimes not resolved from the plasmid DNA fragments of the same molecular weight. To circumvent this problem, we have carried out the experiment shown in Figure 9. Sixteen plasmids which had been classified as "different" by cleavage with various restriction enzymes, were digested with Hha and fractionated on a 1.4% agarose gel. The panel on the left of Figure 9 shows the pattern of digestion products after staining the gel with ethidium bromide. For comparison, the plasmid vector PML-21 was digested with Hha and run in parallel. The DNA fragment which contains the single RI site of this plasmid, and therefore is the fragment into which the chorion sequences are inserted, is indicated by the arrow. As expected, this band is missing from the Hha

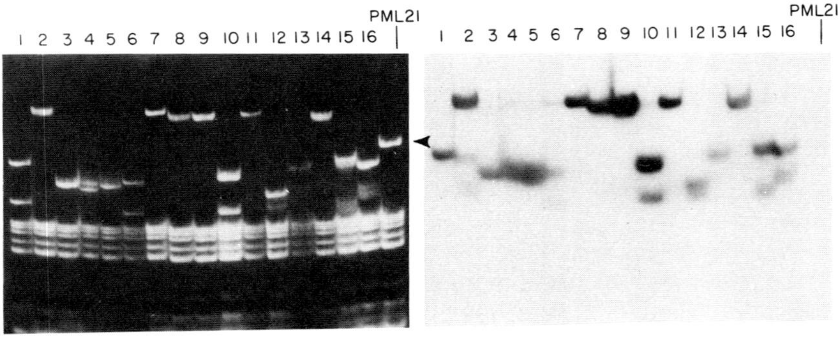

Figure 9. Identification of restriction fragments containing chorion sequences.

Independently isolated hybrid plasmid DNA molecules shown to hybridize chorion cDNA and found to differ in their susceptibility to a large number of restriction endonucleases were digested with the enzyme HhaI, fractionated by electrophoresis on a 1.4% agarose gel and transferred directly to nitrocellulose paper according to the method of Southern (41). The filter was prepared for DNA-DNA hybridization by the method of Denhardt (49) hybridized with ^{32}P-labeled cDNA. and washed as described by Botchan et al. (56). At the left is shown the ethidium bromide stained agarose gel. To the right is the autoradiogram of the nitrocellulose filter. PML-21 designates the gel slot containing the plasmid vector DNA. The numbers at the top of each gel slot indicate the number assigned to the various hybrid plasmids.

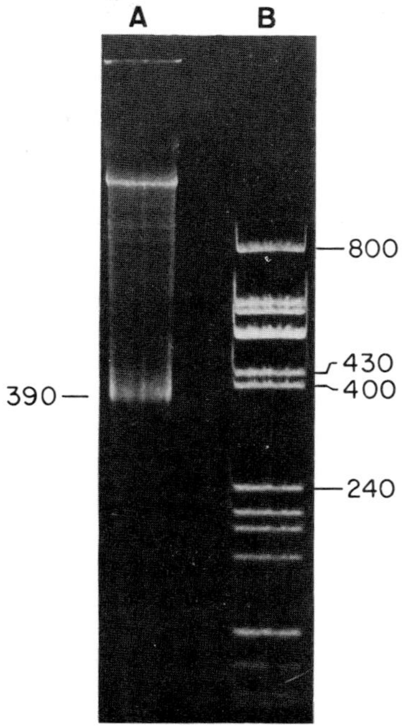

Figure 10. Excision of chorion gene insertion from hybrid plasmid DNA by S1 nuclease digestion.

5 ug of purified hybrid plasmid DNA was labeled in vitro by nick translation using the conditions of Maniatis et al. (47), phenol extracted, ethanol precipitated, resuspended and subjected to S1 nuclease digestion in the presence of formamide as described by Hofstetter et al. (48) and the chorion DNA by electrophoresis on a 3.5% polyacrylamide gel (15). The DNA was visualized by staining the gel with ethidium bromide (0.5 ug/ml) and photographing under UV light. (A) S1 digested plasmid DNA. (B) PMB-9 DNA digested with HaeIII. The size of the markers in base pairs is indicated.

patterns of the hybrid plasmid and is replaced by one or more new fragments depending on whether or not the chorion insertion is cleaved by _Hha_. To determine which of the many bands contain chorion DNA sequence, the DNA in the agarose gel was denatured with alkali _in situ_, neutralized, and transferred directly to a nitrocellulose filter according to the method of Southern (41). As shown in the right-hand panel of figure 9, hybridization with labeled cDNA revealed the restriction fragments which carry chorion DNA sequences. With _Hha_ only one plasmid has chorion DNA fragments which cannot be identified by examination of ethidium bromide gels. However, with many other enzymes such as _Alu_ and _MboII_, several small chorion containing fragments can be distinguished only on the basis of "blotting" experiments. Although many of the plasmids in Figure 9 appear to be similar on the basis of their _Hha_ pattern, for example plasmid DNAs in slots 7,8,9,11 and 14, digestion with different restriction endonucleases indicate that all of the plasmids on the gel are different. It is not surprising that a number of plasmids contain insertions which appear to be similar on the basis of cleavage with one enzyme, considering the extensive homology which has been noted between different chorion proteins. In fact, this is the kind of pattern that one would expect from individual members of multi-gene families.

The presence of sequence homology between apparently different chorion gene plasmids can be demonstrated by cross hybridization between cloned sequences. The relatively low melting temperature of the dA-dT sequences which join chorion and plasmid DNA make it possible to excise the chorion sequence by selectively digesting the A-T linker with S1 nuclease under partially denaturing conditions. Figure 10 shows the S1 nuclease digestion products of hybrid plasmid DNA fractionated on a 3.5% polyacrylamide gel. The chorion DNA insertion of approximately 390 base pairs is easily separated from the bulk of undigested plasmid DNA. To examine the extent of homology between different chorion cDNA plasmids, the same group of plasmid DNAs shown in Figure 9 were digested with _Hha_, fractionated and transferred to a nitrocellulose filter as described above, and _in vitro_ labeled, excised chorion-DNA insertion from plasmid number 10 used as a hybridization probe. To minimize the cross hybridization between partially homologous plasmids, the filters were washed under unusually stringent conditions after the hybridization was finished. As shown in Figure 11, the excised chorion DNA from clone 10 hybridizes to clone 10 DNA as expected, but in addition, hybridizes to four other plasmid DNA molecules (clones 8,17,9,6). Significant hybridization to the other 11 chorion gene plasmids

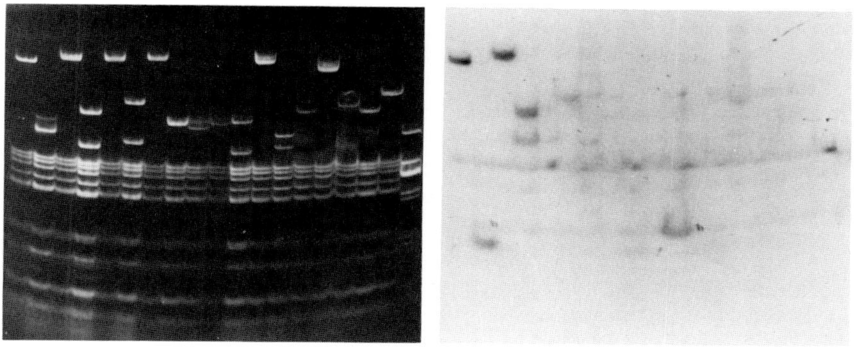

Figure 11. Classification of chorion gene plasmids by hybridization to excised chorion insertions.

Nitrocellulose filters were prepared as in Figure 13 and challenged with *in vitro* labeled, chorion DNA excised from the plasmid designated 10 as described in Figure 12. Hybridization was carried out in 4X SSC, 65°C then washed extensively with 2 x SSC, 30% formamide at 65°C. PMB-9 DNA digested with *Hae*III was included as a molecular weight marker.

on the filter is not observed with these conditions. Although it is possible that the chorion sequences from clones 8 and 9 are simply shorter versions of the insertion in plasmid 10, plasmid 17 and 6 are clearly different because of the difference in the size of the small internal fragments seen in the two plasmids. It is interesting to note that in the case of plasmid 6, hybridization with cDNA (see Figure 9) reveals three DNA fragments which show homology, presumably an internal fragment plus the two flanking fragments containing plasmid sequences. However, as shown in Figure 11, when the excised chorion sequence from clone 10 is used as hybridization probe, only the internal fragment of plasmid 6 hybridizes, indicating that the DNAs from clones 6 and 10 share a common region of very close homology in addition to regions which are nonhomologous. This suggests that a duplication of a small region of one gene has been inserted into another during evolution. Determination of the nucleotide sequences of chorion DNA insertions such as those of clones 6 and 10 should reveal interesting features related to the evolution of multi-gene families (36).

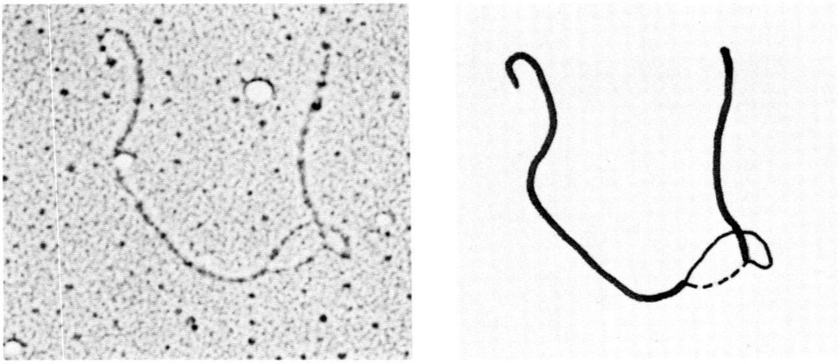

A B

Figure 12. An electron micrograph of a heteroduplex between two independently isolated chorion gene-plasmid DNA hybrids.

Two hybrid DNA molecules isolated from different bacterial clones were digested with the restriction enzyme BglI, which does not cleave within either of the the chorion DNA insertions. Following phenol and ether extractions, aliquots of the two samples were mixed, heteroduplexes prepared and spread for electron microscopy according to the procedure of Davis et al. (46). (A) Photomicrograph showing the position of single stranded regions of mismatch between the two non-homologous chorion DNA insertions. (B) Diagramatic representation of (A) showing the unequal lengths of the single stranded chorion DNA from the two different plasmids. The thick lines represent the duplex regions of the plasmid vector DNA, the thin solid line represents the chorion sequence from one plasmid, while the dashed line indicates the shorter chorion sequence from the other plasmid .

(2) Heteroduplex analysis: The other method of distinguishing between different chorion gene plasmids is heteroduplex analysis. An example of this approach is shown in Figure 12. Two plasmid DNAs which were shown to be different by restriction endonuclease cleavage analysis were linearized by digestion with the enzyme BglI and a heteroduplex between the two molecules prepared. The two chorion insertions appear to be different on the basis of their lack of homology. The mismatched DNA is longer in one plasmid than the other, indicating that the size of the two insertions is different. This approach suffers from the problem that the same chorion sequence inserted in opposite orientations will be scored as different. However, the orientation can

be readily determined using published methods (15).

Although it is difficult to rigorously establish the uniqueness of independently isolated chorion cDNA clones on the basis of only one experimental approach, we have found that by using several different techniques unique cDNA clones can be unambiguously identified. We are presently attempting to classify these unique cDNA plasmids on the basis of their ability to hybridize in vitro labeled, stage specific nuclear and polyA containing cytoplasmic RNA. This will provide a collection of stage specific gene probes which can be used to identify and isolate the DNA sequences which flank individual chorion genes in chromosomal DNA. Comparison of structure of the DNA at the 5' ends of coordinately expressed chorion genes should provide valuable information regarding the mechanism of the control of gene expression in the silk moth.

Acknowledgement

We are grateful to R.J. Roberts for kindly providing many of the restriction enzymes used in this study and to M. Botchan for advice on filter hybridization experiments. We thank M. Moschitta for expert secretarial assistance and B. Yaffe and M. Koehler for preparing the figures. This work was supported by an NSF grant to F.C.K. and T.M.

References

1. Brown, D. and Stern, R. (1974) Annual Review of Biochemistry 43, 667.

2. Birnsteil, M.L., Wallace, H., Sirlin, J., Fischberg, M. (1966) Nat. Cancer Inst. Monogr. 237, 431.

3. Dawid, I.B., Brown, D.D., Reeder, R.H. (1970) J. Mol. Biol. 51, 341.

4. Kedes, L.H. and Birnsteil, M. (1971) Nature New Biol. 230, 165.

5. Kedes, L.H. (1976) Cell 8, 321.

6. Carroll, D. and D.D. Brown (1976) Cell 7, 467.

7. Cohn, R.H., Lowry, J.C. and Kedes, L.H. (1976) Cell 9, 147.

8. Clarkson, S.G., Smith, H., Shaffner, W., Gross, K. and Birnsteil, M. (1976) Nuc. Acids Res. 3, 2617.

9. Wellauer, P.K., Dawid, I.B., Brown, D.D. and Reeder, R.H. (1976) J. Mol. Biol. 105, 461.

10. Clarke, L. and Carbon, J. (1976) Cell 9, 91.

11. Kramer, R.A., Cameron, J.R. and Davis, R.W. (1976) Cell 8, 227.

12. Grunstein, M. and Hogness, D.S. (1975) Proc. Nat. Acad. Sci. 72, 3961.

13. Efstratiadis, A., Maniatis, T., Kafatos, F.C., Jeffrey, A. and Vournakis, J. (1975) Cell 4, 367.

14. Efstratiadis, A., Kafatos, F.C., Maxam, A.M. and Maniatis, T. (1976) Cell 7, 279.

15. Maniatis, T., Sim, G.K., Efstratiadis, A., Kafatos, F.C. (1976) Cell 8, 163.

16. Rougeon, F., Kourilsky, P. and Mach, B. (1975) Nuc. Acids. Res. 2, 2365.

17. Rabbits, T.H. (1976) Nature 260, 221.

18. Higuchi, R., Paddock, G.V., Wall, R. and Salser, W. (1976) Proc. Nat. Acad. Sci. USA 73, 3146.

19. White, R.L. and Hogness, D. (manuscript in preparation).

20. Thomas, M., White, R.L. and Dennis, R.W. (1976) Proc. Nat. Acad. Sci. 73, 2294.

21. Proudfoot, N.J. and Brownlee, G.G. (1976) Br. Med. Bull. 32, 251.

22. Salser, W. et al. (1976) Fed. Proc. 35, 23.

23. Robertson, H.D., Dickson, E., Model, P. and Presnky, W. (1973) Proc. Nat. Acad. Sci. USA 70, 3260.

24. Marrota, C.A., Forget, B.G., Weissman, S.M., Verma, I.M., McCaffrey, R.P. and Baltimore, D. (1974) Proc. Nat. Acad. Sci. 71, 2300.

25. Maxam, A. and Gilbert, W. (1977) Proc. Nat. Acad. Sci. (in press).

26. Sanger, F. and Coulson, A.R. (1975) J. Mol. Biol. 94, 411.

27. Dayhoff, M.O. (1972) Atlas of Protein Sequence and Structure 5, (Georgetown University, Washington: Natl Biomedical Res. Foundation).

28. Legon, S. (1976) J. Mol. Biol. 106, 37.

29. Lockard, R.E. and RajBhandary, U.L. (1976) Cell 9, 747.

30. Efstratiadis, A., Kafatos, F.C. and Maniatis, T. (1977) Cell (in press).

31. Ling, V. (1972) J. Mol. Biol. 64, 87.

32. Garrick, M.D., Hafner, R., Bricker, J. and Garrick, L.M. (1974) Ann. N.Y. Acad Science 241, 436.

33. Kafatos, F.C. (1976) in Control Mechanisms in Development. R.H. Meints and E. Davies eds., Plenum Publ. Corp., New York, pg 103.

34. Paul, M. and Kafatos, F.C. (1975) Devel. Biol. 42, 141.

35. Regier, J.R. (1975) Ph.D. Thesis, Harvard University.

36. Hood, L., Campbell, J.H. and Elgin, S.C.R. (1975) Ann. Rev. Gen. 9, 305.

37. Gelinas, R.E. and Kafatos, F.C. (1973) Proc. Nat. Acad. Sci. 70, 3764.

38. Hunsley, J., Gelinas, R.E. and Kafatos, F.C. (in preparation).

39. Gelinas, R.E. and Kafatos, F.C. (1977) Devel. Biol. (in press).

40. Vournakis, J.N., Efstratiadis, A. and Kafatos, F.C. (1975) Proc. Nat. Acad. Sci. 72, 2959.

41. Southern, E.M. (1975) J. Mol. Biol. 98, 503.

42. Proudfoot, N.J. (1976a) J. Mol. Biol. 107, 491.

43. Proudfoot, N.J. (1976b) Nuc. Acids Res. 3, 1811.

44. Salser, W., Browne, J., Clarke, P., Heindell, H., Higuchi, R., Paddock, G., Roberts, J., Studnicka, G. and Zakar, P. (1976) In: Prog. Nucl. Acids Res. Mol. Biol. W.E. Cohn (ed) Academic Press, N.Y. 17, 177.

45. Jay, E., Bambara, R., Padmanabhan, R. and Wu, R. (1974) Nuc. Acids Res. 1, 331.

46. Davis, R.W., Simon, M. and Davidson, N. (1971). Methods in Enzymology, L. Grossman and K. Moldave (eds) Vol. XXI, page 413.

47. Maniatis, T., Jeffrey, A. and Kleid, D.G. (1975) Proc. Nat. Acad. Sci. 72, 1184.

48. Hofstetter, H., Schambock, A., Van den Berg, J., Weissman, C. (1976) BBA 454, 587.

49. Denhardt, D. (1966) Biochem. Biophys. Res. Comm. 23, 641.

50. Botchan, M., Topp, W. and Sambrook, J. (1976) Cell 9, 269.

DISCUSSION

W. SALSER: Dr. Maniatis and the chairman have suggested that I mention a similar cloning and sequencing project carried out by Gary Paddock, Shin Lin, Chuck Heindell, Jeff Browne, Russell Higuchi and others in my laboratory. Earlier speakers have mentioned their concern about the fidelity of the cloning process and it is interesting that we have observed one case where a single nucleotide has been deleted in a rabbit beta globin gene sequence carried in a plasmid. This deletion could have occurred in any of the steps involved in making the cDNA gene copy or in the growth of the plasmid in the Rec A^- host. Have you observed any similar suggestions of infidelity?

T. MANIATIS: No.

W. SALSER: The particular "mutation" we have detected shows up very clearly in the data obtained using the Maxam and Gilbert methods since it is a deletion. The most interesting comments about the relationship of the sets of data from our two laboratories will have to come from a detected comparison of the sequences obtained. Since this is the first time I have seen the data obtained by Dr. Maniatis and his collaborators I haven't had a chance to make such a comparison.

T. MANIATIS: As I previously mentioned, our sequence was compared to pyrimidine tract and T1 oligonucleotide data and partial mRNA sequences published by other laboratories. Altogether, approximately 97% of the globin sequence from the plasmid could be checked with independently-derived data. In other words, on the order of only 18 out of 576 positions cannot be verified directly. It therefore seems unlikely that any unnoticed differences between the plasmid and mRNA sequence exist.

F. BLATTNER: I am curious about - since you have verified this sequence in so many ways - what your opinion is of the validity of the Maxam Gilbert technique taken alone -or taken in two directions as a completely self contained method.

T. MANIATIS: I think that the method has been perfected to the extent that you can reliably obtain sequences directly without any further verification. I think many other people have made the same conclusion.

R. WU: Tom, I think your method is very beautiful when one can isolate messenger RNA. I just want to describe a general method for cases where one cannot isolate the mRNA. As long as one knows part of a protein sequence, one can use the genetic code and derive a mRNA sequence. Then, one can chemically synthesize a short DNA primer, complementary to part of the mRNA, and use that as a probe to isolate the desired mRNA. After this step, one can go your route or use other methods to get the desired DNA segment for cloning. Using our new method, Jack Szostak has isolated yeast cytochrome c mRNA and then the cytochrome c gene.

P. SARIN: I wonder if you comment on the extraction procedure you used for the extraction of DNA from the gels and whether your procedure could be utilized for extraction on a preparative scale?

T. MANIATIS: The DNA can be recovered in preparative amounts simply by excising the band and soaking it in buffer. We get on the order of 80% recovery routinely. The DNA eluted from gels can subsequently be labeled with kinase, and it can be nick translated with DNA polymerase I. There are problems with DNA fragments isolated from agarose gels, but various ways have been devised to get around this.

W. SALSER: I have two additional questions. First, your gel electrophoresis patterns show a strong band corresponding to the beta mRNA copies but relatively little material outside that band which could be alpha sequences. Since your results suggest that the main band is nearly all beta sequences I wonder what has become of the alpha sequence in your experiments. In our own lab we found that 9 out of the first 35 clones examined were alpha sequences.

T. MANIATIS: As you know, beta-globin mRNA is approximately 50-60 nucleotides larger than alpha-globin mRNA. The globin DNA fragment we inserted into the plasmid was the highest molecular weight double-stranded cDNA on the acrylamide gel. The resolution on these gels is certainly good enough to easily separate duplex molecules which differ by 50 base pairs. It is therefore not surprising that our plasmids contain only beta-globin DNA.

W. SALSER: Second, it occurs to me that you may have been able to make a detailed comparison of part of our sequence data with your own since we recently sent you the galley proofs of our article in Science. Did you notice any disagreements between your data and the 169 nucleotides we reported there?

T. MANIATIS: In addition to the deletion you have found in your sequence, there are two other differences. One is the result of the known isoleucine valine polymorphism in beta-globin. The other difference is at a position in the coding sequence which does not result in an amino acidchange.

W. SALSER: If our mRNA's differ by the isoleucine-valine polymorphism known to exist in rabbit populations it implies that the two genes we have copied have been evolutionarily distinct for a fairly long period. In that case it seems likely to me that the nucleotide difference which does not result in an amino acid change is likely to be a silent polymorphism rather than a case of infidelity in cloning.

T.W. BORUN: Do you think it's theoretically ever going to be possible to study the transcription and posttranscriptional control sequences of any of these messenger RNA's, using technologies based solely on the messenger RNA themselves?

T. MANIATIS: No. The objective of this method is to generate probes that can then be used to identify and isolate sequences adjacent to genes.

T.W. BORUN: Where? In the fragmented DNA and in the cell itself?

T. MANIATIS: Yes, in chromosomal DNA. By performing so-called shotgun experiments, or enriching for particular sequences as John Morrow has described, it should be possible to accomplish this. In the case of the silk moth, we have inserted high molecular weight chromosomal DNA into plasmids and identified chorion gene containing plasmids by hybridization to *in vitro* labeled mRNA. A similar approach could, of course be taken with globin using globin plasmid DNA as probe. The sequences which flank the gene could be cloned and then studied in detail.

T.W. BORUN: How do you foresee going back and studying such flanking sequences in the genome?

T. MANIATIS: We are starting by trying to derive a detailed restriction map of the sequences which flank the globin gene. We are using a method that Mike Botchan has used to study the organization of SV40 DNA sequences integrated into chromosomal DNA. The method seems to be working for globin in that we can measure the distance from certain restriction sites within the globin gene to sites located in the sequences which are adjacent to the gene in chromosomal DNA. We hope to use this information as a guide to enrichment procedures. The map will also make it possible to determine if the cloned adjacent sequences have rearranged during their propagation in *E. coli*.

T.W. BORUN: Your present work now seems to be a tour de force demonstrating that the technology of making the cDNA and the double stranded DNA etc. all work and allow you to definitively determine that you have the same sequences in the plasmid as in the message. However, when you go to these DNA flanking control sequences that presumably aren't presenting messages, are you going to have to be functioning more or less blindly? Would you suggest that these sequencing techniques are solely a way of checking your basic methods before you get to the point where you won't have any RNA's to help you?

T. MANIATIS: As I said, you can check the adjacent sequences at the level of restriction endonuclease cleavage sites by mapping the chromosomal DNA directly. Obviously there is no way we can determine nucleotide by nucleotide whether the cloned sequences are identical to the sequences in chromosomal DNA.

USE OF AN EK-2 VECTOR FOR THE CLONING OF DNA FROM HIGHER ORGANISMS

P. LEDER, D. TIEMEIER, S. TILGHMAN and L. ENQUIST

Laboratory of Molecular Genetics
National Institute of Child Health and Human Development
National Institutes of Health
Bethesda, Maryland 20014

Abstract: We have constructed two strains of bacteriophage λ to meet the requirements of an EK2 host vector system as established by the NIH Advisory Committee on Recombinant DNA Research. Both strains rely on the λgt•λC system devised originally by Davis and his colleagues (1). Here we describe some of the properties of these vectors and their use in cloning a specific <u>Eco</u> R1 fragment of mouse DNA which encodes extensive portions of the mouse ribosomal genes.

INTRODUCTION

We have recently focused our attention on the cloning of DNA fragments from the genomic repetoire of the mouse. While the structural sequences we have heard described in earlier papers at this symposium are of great interest, it is also clear that the answers to questions regarding gene regulation and organization and, as is our concern, the question of antibody diversity will come from understanding the genetic context of the structural sequence.

Obviously, when one works with a mammalian genome, success in cloning a specific DNA fragment is influenced by the enormous size and complexity of the genome. Cloning a given fragment depends upon the power of the available screening technique, the efficiency of the cloning procedure and the purity of the DNA fraction from which the clone is selected. We would like to review briefly a small set of bacteriophage λ vectors which we have modified to produce an EK2 biologic containment system useful in the cloning of DNA from higher organisms (2, 3). We would also like to describe our experience with a high capacity DNA purification technique which we find most promising and which has allowed us to clone a segment of the ribosomal genes of the mouse. Finally, we would also like to share with you some questions which arise when we consider what can be done with this surprisingly versatile hybrid containing the mouse ribosomal fragment.

THE VECTOR SYSTEM

When we first considered which of the host vector systems might prove most useful in this work, it seemed that the λgt system developed by Davis and his co-workers (1) offered real advantages in that it provided a positive selection for phage containing a fragment of foreign DNA (Figure 1). This is so because the λgt system contains only two *Eco* R1 sites which divide the phage into three fragments, two rather large outer fragments and a smaller central fragment. Since the phage already contains a large deletion and since the central fragments contain no genes necessary for phage propagation, the two arms provide all the necessary genetic information required for phage growth. However, they do not provide sufficient length of DNA for phage packaging and, therefore, a fragment of DNA must be inserted between them in order to produce viable phage particles. Theoretically, fragments from 1 to 14 kilobases in length may be inserted (Figure 2). While λ phage itself already possess a number of features which offer advantages as a biological containment system, we introduced three amber mutations in the *W*, *E* and *S* genes which inactivate functions necessary for phage assembly, phage coat protein synthesis and host lysis. These mutations render the phage unable to propagate in any but very specific suppressor containing host cells.

We have performed a large number of tests, including animal studies to demonstrate the validity of these phage as an EK2 host vector system (4, 5). On the basis of these tests, our initial vector, λgt*WES*·λC was certified as an EK2 vector in January of 1976. A derivative phage, λgt*WES*·λB, which substitutes the phenotypically inert λB fragment for the recombination gene-containing λC fragment, is now under consideration for approval as an EK2 vector. The latter phage has the advantage of not requiring the biochemical deletion of the central fragment in cloning experiments using DNA from higher organisms. λgt*WES*·λB has a further interesting property in that the B fragment contains two closely associated *Sst* I restriction endonuclease sites so that (see figure 1) if the vector is first digested with the endonuclease Sst I and subsequently with the endonuclease Eco R1, a cloning vehicle is provided in which the parental phage type will appear only rarely in a population of authentic hybrid phage.

Inasmuch as all our experience in cloning mouse DNA has been with the certified λgtWES·λC system, it is useful to provide an assessment of its efficiency. Theoretically we expect that Eco R1 fragments ranging in size from 1 to 14 kb can be inserted into this vector. Our experience is that approximately 5,000 clones can be obtained from one μg of Eco R1 digested foreign DNA. According to the conditions of certification of this vector system we are required to purify the outer fragments of DNA from the central λC fragment, so this procedure virtually eliminates parental type phage.

THE SCREENING TECHNIQUE

While at this session we have heard Dr. Davis (6) describe an extremely powerful screening technique capable of testing virtually hundreds of thousands of clones in a single day, our own experience has been with a technique developed earlier in his laboratory by Dr. Richard Kramer (7). Using this *in situ* hybridization technique, we have been able to process and screen on the order of 2,000 to 4,000 clones at one time. This should be adequate to obtain a given hybrid in terms of the DNA purification we are able to achieve by the technique described below.

PURIFICATION OF MOUSE DNA ECO R1 RESTRICTION FRAGMENTS

While there are a variety of procedures for purifying restriction fragments of DNA, the size of the mouse genome requires that we use a very high capacity technique in order to isolate fragments that constitute a very small fraction of the total DNA. We have found that the RPC-5 column chromatography which was recently employed for DNA purification (8, 9) is extremely useful in this respect. In Figure 3 is shown the result of applying approximately 12 mg of endonuclease Eco R1 digested mouse DNA to such a column. Each of the 40 or so fractions containing significant amounts of DNA was assayed by hybridization to ^{125}I labeled ribosomal RNA. Two peak fractions were detected, one a relatively homogeneous, early eluting peak and a second, relatively heterogeneous, late eluting peak. When each of these fractions was analyzed by agarose gel electrophoresis

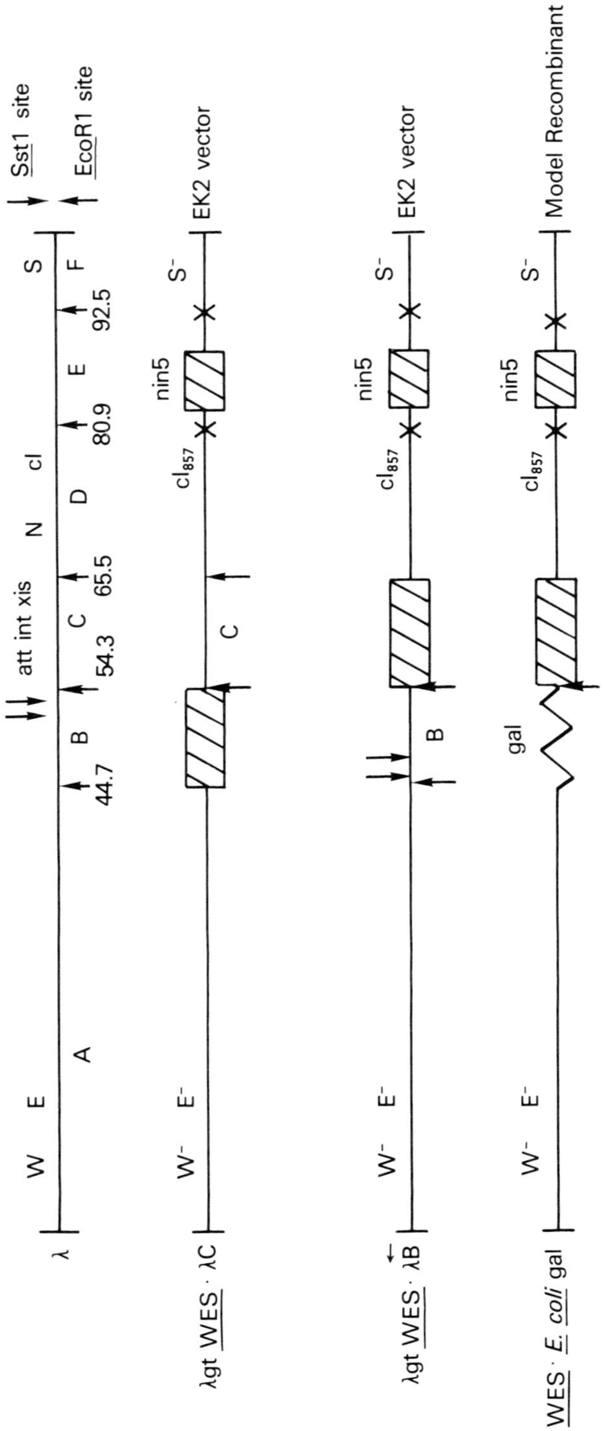
λ
W
E
A
44.7
B
att int xis
54.3
C
65.5
N
D
cI
80.9
E
92.5
F
S
Sst1 site
EcoR1 site
λgt WES · λC
W⁻
E⁻
C
cI857
nin5
S⁻
EK2 vector
λgt WES · λB
W⁻
E⁻
B
cI857
nin5
S⁻
EK2 vector
λgt WES · E. coli gal
W⁻
E⁻
gal
cI857
nin5
S⁻
Model Recombinant

Fig. 1. Modified λ phage suitable for cloning DNA from higher organisms. The lines represent the genome of phage λ. The length of the line drawn represents the full length of the genome of wild-type bacteriophage λ. Letters over each line refer to specific λ genes. Letters under each line refer to Eco R1 restriction fragments of λ with each arrow indicating an Eco R1 site.. The numbers under each arrow represent the position of the site as a percent of the λ genome. Arrows over each line indicate the position of an Sst I site (note inversion as compared to wild type). Scored boxes represent deleted portions of the λ genome. X represents the point at which an Eco R1 site has been eliminated by mutation. The broken line represents the location of E. coli DNA containing the galactose operon. The model recombinant was constructed by crossing λgt Sam100·λC with λ Wam403 Eam1100 gal8 bio256 imm434 cIts1 as described by Enquist et al. (2). The resultant phage eliminated the R1 site at 54.3% λ and substituted an R1 site within and close to the right end of the gal8 insertion. The internal Eco R1 fragment containing a small portion of the gal8 insertion, the λ genes int, xis and a portion of the red gene were deleted by in vitro recombination yielding the phage depicted in the diagram. The position of the left end of the gal8 substitution is at 44.7% on the λ map (W. Szybalski, personal communication), very close to the original λ Eco R1 site.

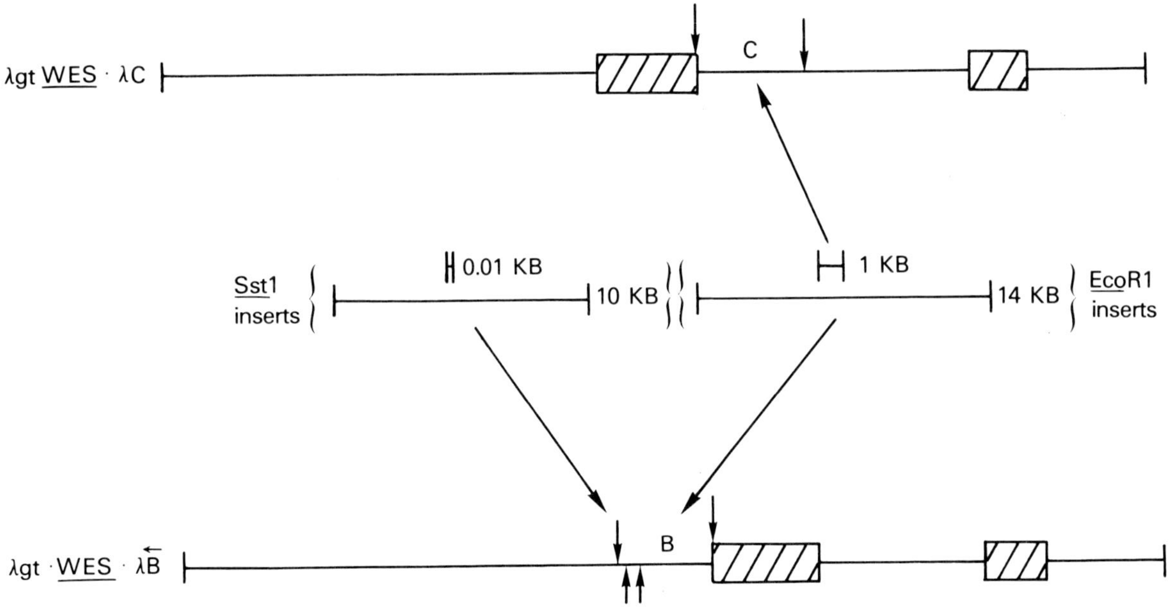

Fig. 2. Length of Eco R1 and Sst I restriction fragments which may be inserted into λgtWES·λC and λgtWES·λB. Symbols are as indicated in the legend to Figure 1. The arrows above the line represent Eco R1 sites, and those below the line represent Sst I sites. Kb represents 1,000 base pairs.

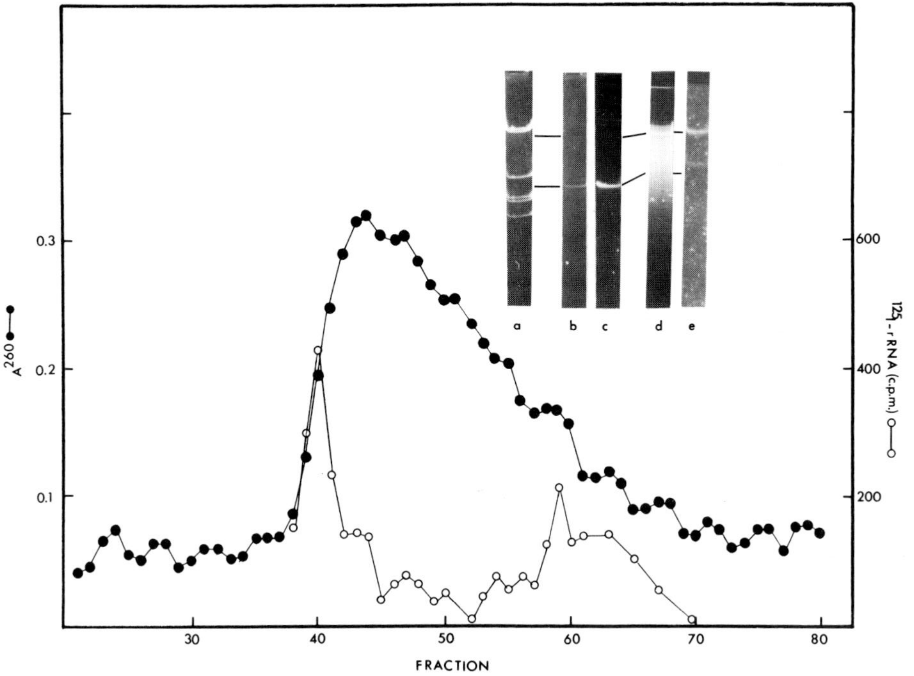

Fig. 3. RPC-5 Chromatography of Eco R1 Fragments of Mouse Genomic DNA and Identification of Fragments Encoding Ribosomal RNA. An Eco R1 digestion of mouse genomic DNA was chromatographed on an RPC-5 column as described in reference 11. ●—● A_{260}; o—o hybridization of ^{125}I-rRNA to individual fractions. Insert: agarose gel electrophoresis of an Eco R1 digest of wild-type λ cI857 DNA where migration is from top to bottom and the kilobase pair sizes of individual fragments are: 21.3, 7.36, 5.79, 5.4, 4.69 and 3.3 kb (lane a); ethidium bromide stain of DNA from fractions 40 and 59 (lanes b and d, respectively); autoradiogram of a Southern transfer of lanes b and d (lanes c and e, respectively), hybridized to ^{125}I-rRNA.

and stained with ethidium bromide, two major bands of DNA appeared in a relatively heterogeneous background of contaminating DNA sequences. These bands could be identified as ribosomal sequences by use of the in situ hybridization technique developed by Southern (10) and discrete bands corresponding to ribosomal gene sequences were identified in this way. A single passage through this column resulted in an approximately 30-fold purification of these ribosomal gene sequences and provided a means for quickly assessing the usefulness and efficiency of the cloning procedure.

CLONING MOUSE RIBOSOMAL GENE SEQUENCE

Recombinant phage were prepared using the isolated arms of λgtWES·λC and the initial peak containing a ribosomal gene fragment 6.6 kb in length. Of the approximately 180 plaques which were picked for screening by in situ hybridization, two were positive and selected for further characterization. Both hybrids later proved to contain the same mouse ribosomal DNA sequences inserted in the same orientation. As shown in Figure 4, the hybrid phage contained a 6.6 kb fragment which comigrated with the prominent fragment seen in the enriched mouse genomic DNA. This fragment was hybridized using the blotting technique to ^{125}I rRNA and revealed a single band at the 6.6 kb position. In addition, if the hybrid containing λ phage was labeled with P^{32} using DNA polymerase I and itself used as a probe against an agarose gel blot of the purified mouse DNA fragment, it also hybridized to a 6.6 kb long DNA band.

The orientation of the mouse ribosomal fragment with respect to the right and left arms of λ could readily be determined from the digestion patterns obtained with the endonucleases Eco R1 and Sal I. As shown in Figure 5, an endonuclease Sal I digestion of the hybrid vector yields three bands, two of which hybridize to total ribosomal RNA. These bands correspond to the left arm of λ plus approximately two-thirds of the insert and a smaller, fragment containing the right-hand third of the insert and a portion of the right λ arm. Hybridization with purified 28S ribosomal RNA, however, was restricted to the longer left-hand fragment indicating that the 28S sequences are located on this fragment and, therefore, the 18S sequence must be to its right, on the smaller Sal I fragment. More detailed studies,

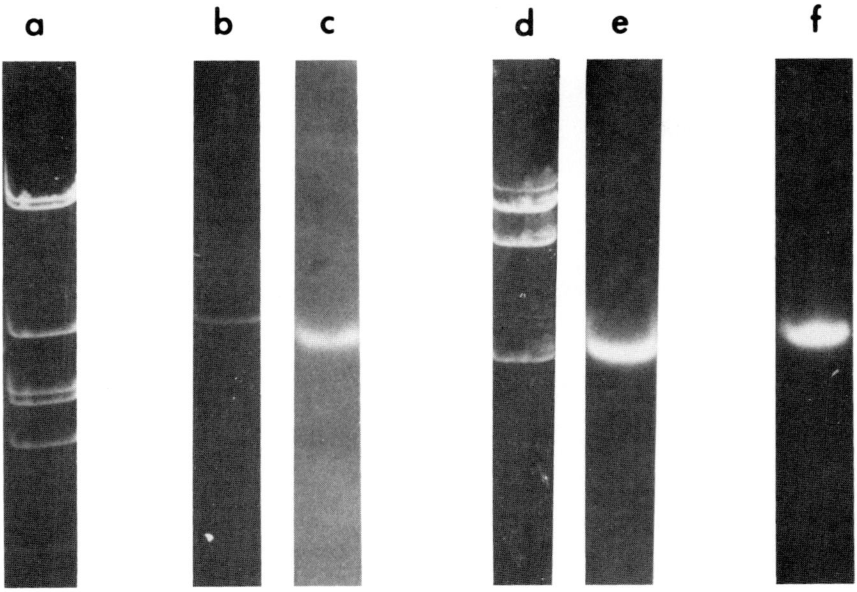

Fig. 4. Eco R1 Digestion of ribosomal DNA and the Hybrid Clone. Ribosomal DNA purified on RPC-5 columns (Fig. 3, fraction 40; lane b) and an Eco R1 digest of hybrid clone A22A2 (lane d) were electrophoresed on a 1% agarose gel. The DNAs were transferred to Millipore filters and hybridized to ^{125}I rRNA using the Southern technique (10) (lanes c and e). Hybrid DNA was labeled using E. coli DNA polymerase I and used to challenge DNA from the purified fraction (lane f). Lane a contains an Eco R1 digest of wild-type λ as described in the legend to Fig. 3. Method details are given in ref. 11.

including R-loop mapping of the 28S region (not shown), confirm this orientation and indicate that virtually the entire 28S sequence is contained within this fragment (11). Furthermore, arguing from what is known about the organization of and direction of transcription of other ribosomal genes [(i.e., 18S → 28S (12, 13)], we can surmise in that cloned fragment, the 18S and 28S RNA sequences are separated by a transcribed spacer region. If this is correct, it further follows that the sense strand of the insert is under the control of the leftward promoter of λ as indicated diagrammatically in Figure 6. That is also the orientation of a chicken ribosomal gene fragment recently cloned in this same vector system (14).

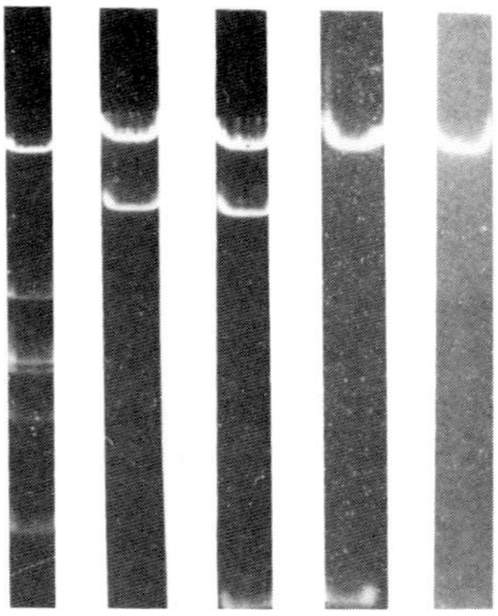

Fig. 5. Sal I Digestion of λgtWES·λC DNA and rDNA Hybrid Clone DNA. DNA from λgtWES·λC (lane b) and the rDNA hybrid clone (lane c) were cleaved with Sal I and electrophoresed on 1% agarose gels. Lane a contains an Eco R1 digest of wild-type λ as described in the legend to Fig. 3. The hybrid clone DNA in lane c was transferred to Millipore filters and hybridized against ^{125}I 18S + 28S rRNA (lane d) or ^{125}I 28S rRNA (lane e). In the diagrammatic representation at the bottom of the figure, the dark arrows indicate the sites of Sal I cleavage in the hybrid DNA. The solid lines represent Sal I fragments which hybridized to the indicated RNA probes.

Figure 6 also indicates diagrammatically the position of a variety of restriction endonuclease sites which occur in this fragment (11). Most interesting among them is the position of two Sst I sites close to the left Eco R1 cut and a Sal I site distal to them. The hybrid phage contains no other Sst I sites, but one other Sal I site which is located in the right arm of λ, approximately 1.4 kb from the right-hand Eco R1 site. This being the case, if the ribosomal fragment were inverted, the Sal I site of the fragment would

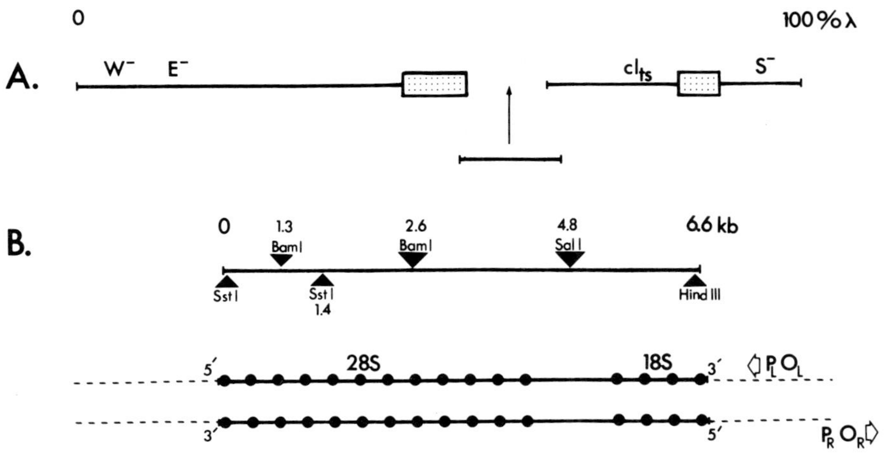

Fig. 6. Diagrammatic Representation of the Orientation of the Ribosomal DNA in Phage λ and the Position of Restriction Endonuclease Cleavage Sites. A. Representation of λgtWES·λC and the position of the inserted ribosomal DNA fragment. B. A restriction map of the inserted fragment, where the numbers above the arrows indicate the distance in kilobases from the left hand end of the insert. In the lower figure, the approximate positions of 28S and 18S rRNA coding sequences are shown, along with the 5'-3' orientation. P_LO_L and P_RO_R represent the approximate positions and orientations of the λ leftward and rightward promoters, respectively.

be located approximately 6.2 kb from the Sal I site in the right arm of phage λ. The two Sst sites would reside between them. Such a phage would be an ideal vector for Sal I fragments inasmuch as removal of this central 6.2 kb fragment would render the arms of insufficient length for phage packaging and the two Sst sites would make the central fragment vulnerable to Sst I digestion. These two properties would, on the one hand, ensure an incorporated fragment in viable phage and, on the other, virtually eliminate parental types among the recombinants.

In summary then, it is clear that the EK2 vector system λgtWES·λC and its derivative, λgtWES·λB, offer sufficient efficiency for the cloning of unique genetic sequences from the DNA of higher organisms. An Eco R1 fragment represented as one of approximately 1,000 fragments is easily within range of the purification screening and cloning procedure. Our own experience with RPC-5 chromatography indicates that several passages over this column achieve this thousand fold degree of purity. Other dimensions of purification would obviously add enormously to the purity. Despite the adverse affect of the NIH Guidelines on the versatility of recombinant DNA technology, it nevertheless appears that the cloning of unique mammalian sequences is within reach.

ACKNOWLEDGMENTS

We are very grateful to Dr. Ronald W. Davis for providing us with the original λgt·λC strain. We are also grateful to Drs. Nat Sternberg, Robert Weisberg and Jacob V. Maizel, Jr. for their helpful suggestions and encouragement. And we would also like to thank Ms. Catherine Kunkle for her exceptional assistance in the preparation of this manuscript.

REFERENCES

(1) M. Thomas, J.R. Cameron and R.W. Davis, Proc. Nat. Acad. Sci. USA, 71 (1974) 4579-4583.

(2) L. Enquist, D. Tiemeier, P. Leder, R. Weisberg and N. Sternberg, Nature, 259 (1976) 596-598.

(3) D. Tiemeier, L. Enquist, P. Leder, Nature, 263 (1976) 526-527.

(4) P. Leder, L. Enquist and D. Tiemeier, Applications to NIH Advisory Committee on Recombinant DNA Research for Certification of λgtWES·λC and λgtWES·λB - December 1976-January, 1977.

(5) P. Leder, D. Tiemeier, L. Enquist, Science, in press.

(6) R.W. Davis, M. Thomas, T. St. John, K. Struhl, J. Cameron, P. Philippsen, R. Kramer and D. Benton, 9th Miami Winter Symposium, 1977 this volume.

(7) R.A. Kramer, J.R. Cameron and R.W. Davis, Cell, 8 (1976) 227-232.

(8) S.C. Hardies and R.D. Wells, Proc. Nat. Acad. Sci. USA, 73 (1976) 3117-3121.

(9) A. Landy, C. Foeller, R. Reszelbach and B. Dudock, Nucl. Acid Res., 3 (1976) 2575-2592.

(10) E.M. Southern, J. Mol. Biol., 98 (1975) 503-517.

(11) D. Tiemeier, S. Tilghman and P. Leder, Cell, manuscript submitted.

(12) P.K. Wellauer, R.H. Reeder, D. Carroll, D.D. Brown, A. Deutch, T. Higachinakagawa and I.B. Dawid, Proc. Nat. Acad. Sci. USA, 71 (1974) 2823-2827.

(13) P.K. Wellauer, I.B. Dawid, D.D. Brown and R.H. Reeder, J. Mol. Biol., 105 (1976) 461-486.

(14) W. McClements and A.M. Skalka, Science (1977) in press.

DISCUSSION

J.T. LIS: Have you any indication of heterogeneity in the repeating units of the ribosomal genes in mouse?

P. LEDER: The only indication, or suggestion that we had was the behavior of the larger fragment on the RBC column which was non-uniform. There is no suggestion of heterogeneity, however, in the cloned fragment when we went back and looked for it in the gene.

F. BLATTNER: What sequence does Sst enzyme cut?

P. LEDER: Rich Robertson tells me it makes a staggered cut and produces large fragments.

F. BLATTNER: Do you know whether it has a 3'- or 5'-extention. Do you try to do tailing with it?

P. LEDER: No. We tried one unsuccessfull experiment with tailing.

J.F. MORROW: If it were possible to get it certified could you grow your phage vectors in a rec^- strain?

P. LEDER: The phage can be grown in a rec^- strain as far as I know. Transfection efficiency has been lower in rec^- strains than in the strains we use (derived from 803).

J.F. MORROW: Are your phage reccombinants rec^-?

P. LEDER: Yes.

J.F. MORROW: Was the certified host rec^+?

P. LEDER: The host that is being considered for use with lambda phage for propagative growth is rec^+.

DNA CLONING IN BACTERIA AS A TOOL FOR STUDY OF IMMUNOGLOBULIN GENES

B.MACH, F.ROUGEON, S.LONGACRE and M-F.AELLEN
Department of Pathology
University of Geneva Medical School
I2II Geneva, Switzerland

Abstract: It has recently become possible to synthesize in vitro double stranded DNA from specific mammalian mRNA templates and to introduce these gene sequences into bacterial plasmids(I).Such plasmids have been constructed containing α and β globin gene sequences from rabbit and mouse and immunoglobulin light chain sequences from mouse. These plasmids are now used for the purification of the mRNA and cDNA probes necessary in the study of immunoglobulin genes. They have also been used for the simultaneous purification of the + and the - strands of cellular DNA restriction fragments carrying the light chain genes.

INTRODUCTION

The elucidation of the mechanisms responsible for the genetic control of the diversity of antibodies still represent a great challenge.The two theories proposed to explain the generation of antibody diversity,the germ line theory of inherited diversity and the theories implying a somatic aquisition of diversity,have opposed immunologists for sometime. The possibility of isolating the mRNA for immunoglobulin light chains (2) had allowed a new approach based on an estimation of the number of genes for immunoglobulin in cellular DNA. In these experiments,we have used either ^{3}H-cDNA or ^{125}I-mRNA as probe,and followed their hybridization to a large excess of cellular DNA (3,4,5).It was concluded from these experiments that the number of genes detectable was low,probably between one and five only.The findings were considered to be strongly in favor of a somatic origin for diversity.Similar observations were made in several other laboratories.

Because of the limitation in the precision and in the interpretation of such type of studies, it was felt that a more direct approach was necessary,with the emphasis no longer on the number of immunoglobulin genes,but rather on their structure.Ultimately, this could lead to a direct demonstration of a somatic change in the nucleotide sequence of V (variable) genes and perhaps to an understanding of the nature of such changes and of the mechanisms responsible. The obvious requirements for such direct approach,ie. pure probes and purified genes,suggested that molecular cloning of DNA in bacteria could be a decisive tool in this field. As a first step,we have developed a procedure for the synthesis of a specific gene sequence from mRNA,followed by the insertion of that sequence into a plasmid of E.coli. With this mRNA-directed approach,it is possible to envisage the insertion of purified or semi-purified gene sequences,which is a much more favorable situation than when total cellular DNA of very high complexity is used as starting material.With rabbit globin mRNA,this procedure led to the first construction of bacterial recombinants carrying a messenger RNA-derived gene sequence (I).In this paper we will review some more recent results on the construction of plasmid recombinants containing globin and immunoglobulin light chain sequences (Fig.I) and discuss some of the possible uses of "light chain plasmids" in the study of immunoglobulin genes.

RESULTS AND DISCUSSION

Synthesis of double stranded DNA

The in vitro synthesis of double stranded DNA from mRNA-derived cDNA was studied initially with rabbit and mouse globin mRNA,which are easily available in large amounts.In the first experiments,the synthesis of the second strand was directed by an oligo-dA primer (and the cDNA template elongated with dT) (1).It was then observed that globin cDNA could serve as a "primer-template" and that AMV DNA polymerase could synthesize the second strand of globin cDNA in a "self-priming"

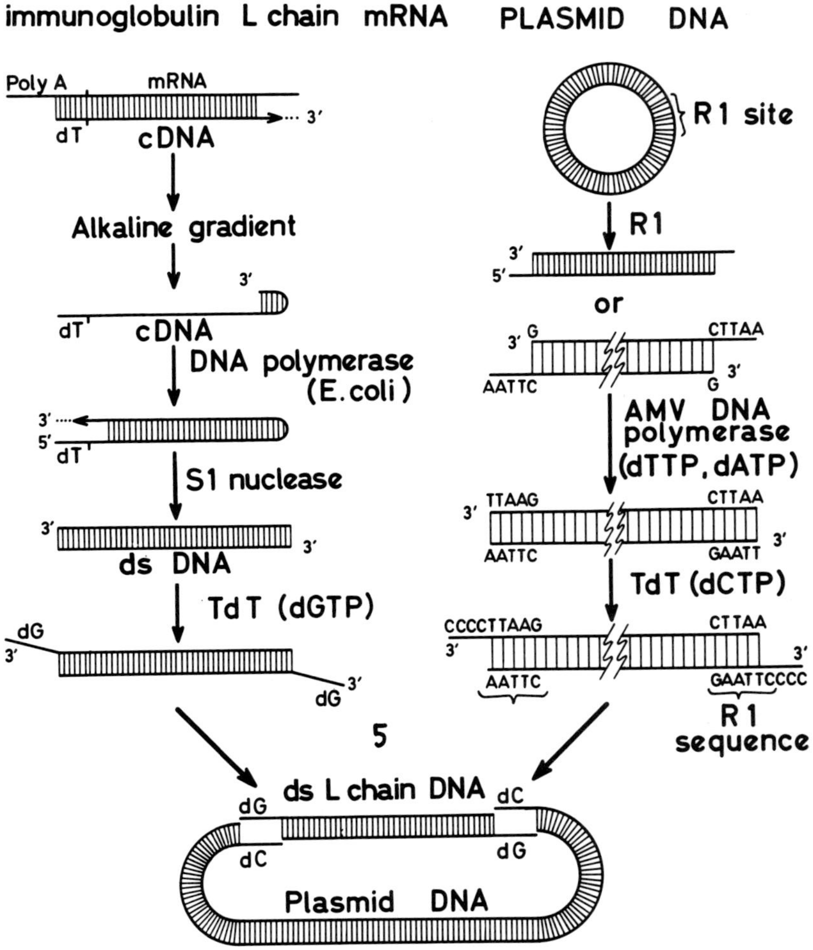

Fig. I. Schematic representation of the general procedure used for the in vitro synthesis of double stranded DNA from mRNA and for the construction of bacterial recombinant plasmids containing these DNA sequences.

reaction (6). The same reaction was described by others with E.coli DNA polymerase (7,8). When immunoglobulin light chain mRNA prepared from mouse plasmocytoma (4) was used as template, it was observed that the cDNA could be copied efficiently by DNA polymerase from E.coli but not by AMV enzyme(6). With the use of L chain mRNA as template,two other aspects,not evident in the case of globin mRNA,had to be dealt with. The purification of the mRNA,which is based on size fractionation,had to be carried through two successive sucrose gradients,(both monitored by the analysis of the translation products on acrylamide gels) followed by preparative acrylamide gel electrophoresis in 98% formamide (4).In favorable cases the homogeneity of the resulting mRNA can be shown by Rot analysis to be almost indistinguishable from that of globin mRNA (see Figure 2), although the degree of purification varies in different preparations.This absolute requirement for a preparative gel fractionation implies important losses.Only very small amounts of L chain mRNA is thus available.The second limitation of this system is that only a small percentage of the cDNA molecules synthesized represent the full length transcript of the mRNA,even when high concentration of triphosphates are used (9). Full-length complement of the mRNA can however be purified by fractionation on sucrose gradients or acrylamide gels.

Construction of recombinant plasmids.

The principle of the insertion of mRNA-derived DNA sequences into E.coli plasmid pCR1 is illustrated in Fig.1.It has been used to construct plasmids with rabbit and mouse globin sequences and mouse immunoglobulin sequences (1,10,11). Other laboratories have used a similar procedure to construct recombinant plasmids containing mRNA-derived sequences (12,8). After elongation of the specific DNA with dG and of linear plasmid DNA with dC, the two DNAs were annealed and the hybrid used for the transformation of $CaCl_2$ sensitized E.coli cells to Kanamycin resistance (1).Circularization of linear plasmids can only take place through the insertion of one (and only one)globin (or immunoglobulin) DNA segment,which should provide an absolute positive selection for recombinants. However,we

have observed a significant number of Kanamycin resistant colonies in control samples with "linear"plasmids alone. This background,which results from traces of remaining circular plasmids(considerably more efficient in transformation), can account for a large proportion of the Kanamycin-resistant clones obtained with globin DNA or L chain DNA hybrids.
The Kanamycin-resistant clones obtained were assayed for the presence of globin or L chain specific sequences by hybridization of ^{3}H-cDNA or ^{32}P cRNA probes,either in liquid or on nitrocellulose filters.In the case of light chain mRNA,the cDNA probe used was first purified by a preparative hybridization to the majority component of the mRNA preparation ("Rot-purified" cDNA) and corresponded therefore only to L chain mRNA sequences.Bacterial transformation and the identification of clones was done in a restricted laboratory, under negative pressure,and with autoclaving of all solution and equipment used. Only 3 such experiments were performed,two with rabbit globin DNA (1,10) and one with mouse globin DNA and L chain DNA (MOPC 173 plasmocytoma) (11,13).

From several clones identified as containing rabbit α or β globin or L chain sequences, DNA was prepared for further analysis of the plasmid and of the inserted sequence. In all cases these studies (assay of hybrids by S1 nuclease and by hydroxylapatite,cross hybridizations,melting temperature) confirmed the specificity of the identifications.In the case of a "light chain plasmid" (pCR1 k 38), a further test was performed after purification of cDNA and of mRNA complementary to the inserted sequence (see next section). The plasmid-purified cDNA was found to correspond indeed to the first kinetic component of L chain mRNA by Rot analysis,and the plasmid-purified mRNA was found to code for immunoglobulin light chain in a wheat-germ system (13).

Size of inserted sequences.

As discussed in detail elsewhere (10,11,14)

the size of the inserted sequences was estimated by hybridization of cDNA probes of known length to plasmid DNA and by electron microscopic studies of heteroduplex structures formed between pCR1 DNA and the DNA of recombinant plasmids (14). In addition, in the case of a rabbit β globin recombinant, in which an EcoRI endonuclease site had been reconstructed on each side of the insert, the inserted gene sequence could be retrieved by EcoRI digestion (10). Because of the presence of an internal EcoRI site within the globin sequence, 2 fragments were recovered, and their sizing (410 and 210 base pairs respectively) provided an additional measurment of the length of the inserted sequence. The results obtained by these different procedures in the case of different plasmids are summarized in Table 1.

Table 1

Estimated length of the specific DNA segments inserted into different recombinant plasmids.

Recombinant plasmid	A Hybridization to ^{3}H-cDNA	B Heteroduplex analysis (EM)	C Size of EcoRI fragments
pCRIβR19	538	620	620
pCRIβR7	540	630	
pCRIαR11	440	485	
pCRIβM9	495	618	
pCRIαM4	540	620	
pCRIk38	810	828	
pCRIk36	290	385	

Sequence length is estimated in base pairs. Plasmids: α and β = α or β globin, R=rabbit, M= mouse, k= mouse kappa light chain. The results of column A (see ref, 10, 11) are derived from the percent protection of long ^{3}H-cDNA (sized by acrylamide gel electrophoresis).The data of column B come from Rochaix et al(14) and are ± about 50 base pairs.For column C,see ref 10.

In the case of the light chain plasmid pCR1 k 38, it was observed that, in addition to an insert of about 820 base pairs,an important deletion had been generated.The size difference on gels,the loss of the SAL I site present in pCR1 and the EM analysis of the heteroduplex formed between pCR1 and pCR1 k 38 indicated that the deletion concerns a sequence of about 3 kilobases, extending from the EcoRI insertion site to the Kanamycin stem of pCR1 (14).

Polarity of the inserted globin or immunoglobulin gene sequence.

It has been determined by Maniatis *et al*(12)that, in a pMB9 plasmid recombinant containing a rabbit β globin gene sequence,the inserted sequence was oriented with the region corresponding to the 3' end of the mRNA closest to the HindIII site. All other β globin recombinants examined by these authors had the same polarity.In the case of one of our pCR1 recombinants carrying a rabbit globin sequence, the same orientation was found (E.Southern and B.Mach,unpublished data) and EM analysis of several heteroduplexes between plasmids carrying rabbit and mouse β globin sequences indicated the same polarity for mouse globin sequences (14). It thus became of interest to examine the orientation of an unrelated sequence (L chain).We took advantage of the important deletion of pCR1 k 38 ,which had shortened the distance between the insert and the Hind III site to approximately about 700-800 base pairs and digested pCR1 k 38 first with Hind III and then with lambda 5' exonuclease.Control and resected DNA were hybridized with ^{3}H-cDNA and with ^{125}I-mRNA,respectively probes for the anticoding and the coding strand. The results indicated (B. Mach,unpublished data) that the coding strand had been resected by λ exonuclease and that therefore the region of the L chain gene corres-

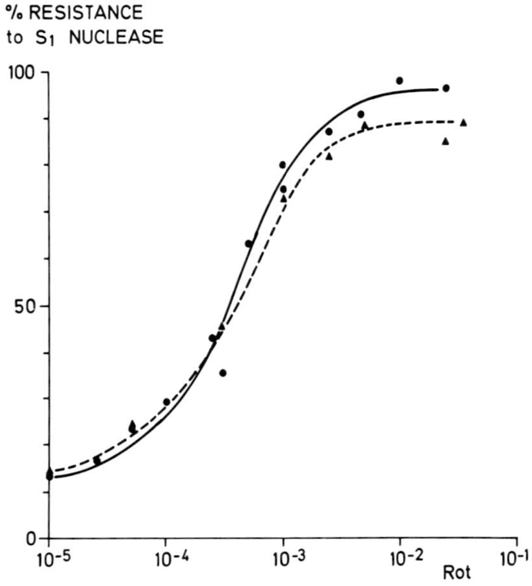

Fig 2. Hybridization kinetics of ^{3}H-cDNA to mRNA.L chain mRNA(MOPC 173) was purified as described (4) including preparative acrylamide gel electrophoresis in formamide. ^{3}H-cDNA was synthesized (3,6) and hybridized to the mRNA at different times in 0.3M NaCl at 68°C.The % cDNA in hybrid form was determined after digestion with nuclease S1 (6).▲-----▲:L chain mRNA and cDNA.●——●: The same experiment with mouse globin mRNA and cDNA (6). Rot is RNA concentration (moles/L) x time of hybridization (sec).

ponding to the 3' end of the mRNA (3' untranslated region:UR, and constant region:C)was oriented towards the Hind III site (Fig 3). Since it is likely that insertion occurs randomly in both orientations,this finding of a unique polarity, if confirmed in other examples,would strongly suggest that the opposite polarity is lethal.

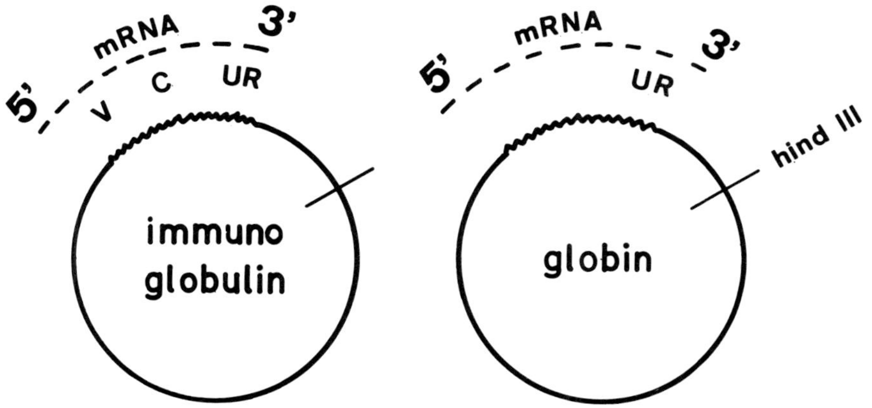

Fig 3. Orientation of immunoglobulin L chain and globin gene sequences carried by recombinant plasmids.The gene sequences had been inserted at the EcoRI site and the Hind III site is indicated as reference.

The possibility of excising the globin sequence of a recombinant plasmid with EcoRI endonuclease allowed the insertion of the excised fragments into lambda bacteriophage and the construction of "lambda-globin" recombinants (15). These were used by Kourilsky et al.(15) to study the expression of globin mRNA in cultures of lysogens made with these lambda recombinants.Use was made of the availability of globin DNA in large excess(in the form of globin-specific plasmid DNA) for the detection of traces of globin RNA in such cultures.In the case of 4 lambda globin recombinants,the data indicate the presence of globin-specific RNA (15).

Use of recombinant plasmid DNA for the purification of cDNA and mRNA.

One of the crucial features in the study of immunoglobulin genes is the need for pure probes. The next step was therefore the use of recombinant plasmid DNA for the purification of the complementary DNA or mRNA.Plasmid DNA could be used for the preparative hybridization of complementary strands either in an insoluble form linked to cellulose or in liquid, following mercuration of the DNA. The hybrids could be recovered by chromatography on columns of SH-Sepharose, according to the principle of Ward (16). We have modified this procedure in introducing, after the binding of the hybrids to SH-Sepharose, a strand separation step (melting in formamide) which allows the recovery of the non-mercurated strand (cDNA or mRNA) (17). The details of the conditions used for mercuration, hybridization, SH-Sepharose chromatography and elution by strand separation will be described elsewhere (17). Using plasmid DNA carrying either a mouse β globin sequence or a mouse L chain sequence, cDNA was purified respectively from mixed α and β globin cDNA, and from cDNA synthesized from crude 14 S plasmocytoma RNA. Table 2 indicates that the cDNA purified is indeed specific to the mercurated plasmid used in the purification. When hybridization is carried to completion, the yield of this purification procedure is between 80-90% of the sequence originally present.

Table 2

Sequence specificity of ^{3}H-cDNA purified with mercurated recombinant plasmid DNA by SH-Sepharose chromatography.

Hg-plasmid used for cDNA purif.		Plasmid DNA used for hybridization with purified cDNA: pCRI	pCRIβM9	pCRIαM4	pCRIk38
pCRI M9 βglobin	S 1:	2.2%	83%	2.5%	-
	HAP:	2.7%	94%	3.2%	-
pCRIk38 Lchain	S 1:	2.1%	-	-	85%
	HAP:	3.1%	-	-	95%

^{3}H-cDNA synthesized from total mouse globin m-RNA (α and β) and from crude 14 S MOPC 173 plasmacytoma mRNA (first sucrose gradient) (2,6) was hybridized to sonicated mercurated plasmid DNA and chromatographed on SH-Sepharose as described elsewhere (17). ^{3}H-cDNA eluted in 97% formamide at 45° C was tested for its complementarity to different recombinant plasmids. The hybrids were assayed either by S1 nuclease digestion or by chromatography on hydroxylapatite (HAP) (1,17).

The use of pure L chain cDNA for the study of the untranslated region of L chain mRNA.

L chain mRNA contains a 200 nucleotide long region on the 3' end, between the poly A and the beginning of the C (constant) region (see diagram of L chain mRNA in ref. 5). The genetic control of the untranslated region and its function are unknown. One important question is whether that sequence is different or conserved among the mRNAs of different immunoglobulin chains (18). To explore this problem, cDNA corresponding to the first 200-300 nucleotides (3'Untranslated Region: 3'UR) of MOPC 173 L chain mRNA was purified with mercurated pCR1 k 38 plasmid DNA and hybridized to mRNA coding for several different immunoglobulin chains. It was observed (18) that lambda chain mRNA did not share that sequence but that all mRNAs of the kappa chain type tested cross-hybridized with MOPC 173 cDNA-3'UR (Table 3). Similar studies with cDNA of other kappa chain mRNAs confirmed this observation. The extent of sequence conservation between the different 3' untranslated regions was further explored with Tm studies and by treatment of the ^{3}H-cDNA:mRNA hybrids with nuclease S1 followed by fractionation and sizing of the S1-resistant cDNA on formamide gels. This later procedure, which provides a stringent test for the quality of hybrids and therefore for sequence conservation and identity, revealed no difference between the 3'UR of the three different kappa chain mRNA studied (18). Finally, the 3'UR region of iodinated and plasmid-purified L chain mRNA was analyzed by a T1 fingerprint and the sequence conservation and

Table 3

Study of the sequence homology in the 3' untranslated region of different mouse light chain mRNAs by hybridization.

		^{3}H cDNA			
		kappa			lambda
		61	173	41	104
mRNA	61:	96%	92%	95%	4%
	173:	97%	95%	94%	4%
	41:	99%	97%	92%	-
	104:	-	-	-	64%
	glo:	1%	1.5%	0.4%	-

L chain mRNA (gel purified, ref. 6) and cDNA of short size were prepared and hybridized as described in detail elsewhere (18). Light chains of MOPC 41, MOPC 173 and MOPC 61 plasmacytomas were of the kappa while that of MOPC 104 is a lambda chain.

probable identity among different kappa chains was confirmed (18). One of the implications of these data is that the 3'UR can be considered genetically as an extension of the kappa light chain C gene. The existence, in mRNA of the lambda chain type, of an untranslated region of the same length but with an unrelated sequence (18) is also of interest.

Purification of immunoglobulin genes from cellular DNA by hybridization to mercurated recombinant DNA

In addition to their use in the purification of specific probes, the recombinant plasmids with globin or L chain gene sequences can be used for the purification of genes from total cellular DNA. It is evident that ultimately the structure of interest is the gene(s) corresponding to the V region in the DNA of germ line cells, prior to a possible somatic acquisition of diversity. Purification of genes from cellular DNA could allow their study in a less complex state than in total DNA, and might also make possible the cloning of these genes in bacterial vectors.

We have approached this problem with the use of mercurated sonicated DNA from globin or L chain specific plasmids which were hybridized to cellular DNA, previously fragmented with restriction endonucleases (EcoRI or BAM I). This procedure, which is outlined in Figure 4, is based on the method developed for the purification of cDNA and mRNA with mercurated plasmid DNA (see above) and which involves melting of the purified strand from the SH-Sepharose column (17). The interest of this procedure is that it allows one to carry the hybridization out in liquid (where the rate of the reaction is faster) and also that it concerns simultaneously the two strands of the specific DNA fragment. A final fractionation of the purified DNA on alkaline sucrose gradient was introduced to eliminate low molecular weight material (resulting from endonucleases) as well as possible traces of sonicated plasmids. The purified DNA was tested by hybridization with ^{3}H-cDNA and ^{125}I-mRNA. Starting with 50 mg of cellular DNA, it was possible to achieve more than a 1000 fold purification ot the specific genes, with about 40% recovery (S. Longacre and B. Mach, unpublished data). A second cycle of hybridization and purification could be performed, with negligible losses of the specific sequences thereby allowing a final gene purification of several thousand fold.

To explore the integrity of the purified DNA sequences, an aliquot of DNA purified from plasmacytoma MOPC 173 with an L chain specific plasmid

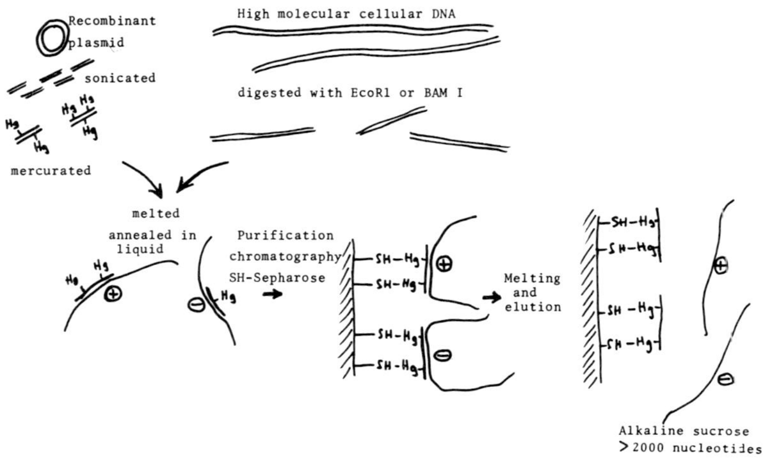

Fig 4. Schematic representation of the procedure used for the purification of cellular DNA fragments carrying globin or immunoglobulin genes by hybridization to mercurated recombinant plasmid DNA (see text).

was chromatographed on an agarose gel under alkaline conditions (19), transferred to a nitrocellulose paper (20), hybridized with plasmid purified ^{125}I-L chain mRNA and analyzed by radioautography. This showed a unique L chain specific DNA band, with a chain length of about 5500-6000 nucleotides (unpublished data). The possibility of purifying with recombinant plasmid DNA both strands of DNA fragments carrying immunoglobulin genes, and the demonstration of the integrity of detectable amounts of the purified fragments suggest that this simple procedure might be a valuble contribution to efforts aimed at the cloning of cellular genes of higher organisms.

REFERENCES

(1) F. Rougeon, P. Kourilsky and B. Mach, Nucl. . Acid Res., 2 (1975) 2365.

(2) B. Mach, C.H. Faust and P. Vassalli, Proc. Natl. Acad. Sci, U.S., 70 (1973) 451.

(3) C.H. Faust, H. Diggelmann and B. Mach, Proc. Natl. Acad. Sci. U.S., 71 (1974) 2491.

(4) M-G. Farace, M-F. Aellen, P-A. Briand, C.H. Faust, P. Vassalli and B. Mach, Proc. Natl. Acad. Sci. U.S., 73 (1976) 727.

(5) B. Mach, M-G. Farace, M-F. Aellen, F. Rougeon and P. Vassalli, Proceedings 10th FEBS Meeting, (North Holland Publ. Co. 1975) p. 299.

(6) F. Rougeon and B. Mach, Proc. Natl. Acad. Sci. U.S., 73 (1976) 3418.

(7) A. Efstratiadis, F.C. Kafatos, A.M. Maxam and T. Maniatis, Cell, 7 (1976) 279.

(8) R. Higuchi, G.V. Paddock, R. Wall, and W. Salser, Proc. Natl. Acad. Sci. U.S., 73 (1976) 3146.

(9) H. Diggelmann, C.H. Faust and B. Mach, Proc. Natl. Acad. Sci. U.S., 70 (1973) 693.

(10) F. Rougeon and B. Mach, J. Biol. Chem. (1977), in press.

(11) F. Rougeon and B. Mach, Gene (1977), in press.

(12) T. Maniatis, S.G. Kee, A. Efstradiadis, and F.C. Kafatos, Cell, 8 (1976) 163.

(13) B. Mach and F. Rougeon, manuscript in preparation.

(14) J-D. Rochaix, F. Rougeon and B. Mach, manuscript in preparation.

(15) P. Kourilsky, D. Gros, F. Rougeon and B.Mach, manuscript in preparation.

(16) R.M.K. Dale and D.C. Ward, Biochemistry, 14 (1975) 2458.

(17) S. Longacre, F. Rougeon, M-F. Aellen and B. Mach, submitted.

(18) M-F. Aellen, P. Vassalli and B. Mach, submitted.

(19) M.W. McDonell, M.N. Simon and F.W. Studier, J. Mol. Biol. (1977), in press.

(20) E. Southern, J.Mol. Biol., 98 (1975) 503.

This work was supported by the Swiss National Science Foundation. P-A. Briand, M. Oligiati and M. Badan provided excellent technical assistance. AMV DNA polymerase was generous gift from Dr. Beard, Viral Cancer Program, Natl. Cancer Institute, U.S.A.

DISCUSSION

P. LEDER: In view of the fact that there is some information available about the distribution of the restriction sites in DNA corresponding to immunoglobulin genes, this experiment can be related to that. In generating plasmids containing immunoglobin sequences, what restriction enzyme was used, what was the source of the DNA, and how long was the fragment?

B. MACH: It was MOPC 173 DNA; this represents the differentiated cell after translocation of the V and C genes. We have had background difficulties looking at V and C genes in total cellular DNA. However, if one is willing to use 50 ml. of DNA and a large amount of enzyme, I am convinced that this procedure will be able to reveal the exact siteThe analysis is more precise than previous approaches.

P. LEDER: Which probe was used, the long one or the short one, for purification of the DNA fragments.

B. MACH: The plasmid used was the one with the long insert.

P.H. ROY: Can the SH-Sepharose column method be used with mercurated RNA as a more general method of pulling out complementary sequences?

B. MACH: We have less experience with mercurated RNA. We have used mercurated DNA to pull out messenger RNA. We are using mercurated mRNA under conditions of R-loop formation as a procedure to enrich specific fragments obtained in the native form. Our results are too preliminary but I think that putting mercury on the messenger and using it as a trap under R-loop conditions is a good approach.

P.H. ROY: Is there a minimum size to the sonicated DNA that will work?

B. MACH: Sonicated DNA was sized on gradients with internal markers. It averaged about 300 to 400 nucleotide pairs.

DNA DEGRADATION BY RAT INTESTINAL NUCLEASES

L. J. Maturin, Sr. and Roy Curtiss III,
Department of Microbiology, University of Alabama in Birmingham, Birmingham, Alabama 35294.

Enfeebled strains of *Escherichia coli* have been constructed for use in safe recombinant DNA research. The first of such strains, χ1776, is unable to survive passage through the intestinal tracts of rats due to a number of genetic safety features including requirements for diaminopimelic acid (DAP) and thymine and sensitivity to bile salts. Because intact or partially degraded DNA may be released upon cell lysis and because little is known about rat intestinal nuclease activity, it is important to determine the "survival potential" (i.e., ability to transform or transfect other bacteria) of the released DNAs. We have extracted DNA from bacterial cells and determined its rate of degradation by rat intestinal contents to be essentially instantaneous in the diluted rat intestinal contents. The above observation coupled with the stringent requirements necessary for transformation or transfection make the mechanism of *in vivo* transmission highly improbable.

Research supported by the National Science Foundation under Grant No. GB-37546.

RESTRICTION ANALYSIS AND CLONING OF DNA FROM THE *ILV* GENE CLUSTER OF *E. COLI* K12

G. M. McCorkle and H. E. Umbarger, Department of Biological Sciences, Purdue University, West Lafayette, Indiana 47907

Analysis of polar Mu-1 insertion mutants has shown that transcription of the *ilvEDA* operon is from *ilvE* to *ilvA* and suggests that specific regulation of the *ilv* cluster is even more complex than previously believed. [J. M. Smith, D. E. Smolin and H. E. Umbarger, Mol. gen. Genet. (1976) in press]. In preparation for *in vitro* analysis of individual regulatory elements, the physical and genetic maps of the *ilv* cluster have been correlated and restriction fragments of the *ilv* region have been cloned and purified.

Cleavage maps of DNA from λcI857St68h80d*ilv* (λh80d*ilv*), containing *ilv* DNA derived from the F16 episome, and of DNA from λp*ilvAD*, containing *ilv* DNA derived directly from the bacterial chromosome, were prepared using *Eco*RI, *Hind*III, *Bam*HI and *Sal*I endonucleases. Comparative analysis shows that λh80d*ilv* contains 15.3 kb of F16 DNA of which at least one-third is present in the normal K12 chromosomal order. λh80d*ilv* restriction fragments, generated by *Hind*III, *Bam*HI and a combination of *Eco*RI and *Hind*III, have been covalently inserted into plasmid pBR313 and used to transform appropriate hosts. Complementation, hybridization, and mapping data show that at least the *ilv* EDAOC region is present in normal chromosomal order on λh80d*ilv*. This region is bounded by a *Bam*HI cleavage site on the *ilvC* side and a *Hind*III cleavage site on the *ilvE* side; it contains an unusual cluster of either four or five *Eco*RI cleavage sites in a 2.8 kb region believed to contain the *ilvAOC* genes. A 5.7 kb fragment of F'DNA also appears to have been incorporated in λh80d*ilv* but the *rrnB* genes from the episome are not present. [See Lee *et al.*, J. Mol. Biol. 89:585 (1974)]. We have not confirmed the presence of chromosomal markers outside the *ilv* cluster reported by Lo Schiavo et al. [J. Mol. Biol. 99:353 (1975)].

The *ilv*-containing recombinant plasmids have been amplified in *E. coli* cells and their DNA purified for *in vitro* analysis; the latter work is now in progress.

THE LOCALIZATION OF THE ECO R_1-SENSITIVE SITES ON THE CHROMOSOMES OF DIFFERENT BACTERIA

A. Prozorov, N. Kalinina, L. Gening, I. Tichonova, V. Debabov
Institut of General Genetics, Moscow
Institut of Industrial Microorganisms Genetic, Moscow

The transforming activity of DNA from Bacillus subtilis 168, B. subtilis W23, B. subtilis NRS, B. aterrimus, and E. coli after the Eco R1 restrictase treatment was studied.

In the case of Bacilli, the different auxotrophic mutants of B. subtilis 168 were used as recipients in transformation experiments. After treating with Eco R1, the transforming activity of the markers were reduced, but to different degrees. With homologous transformation, almost absolute loss of biological activity was observed in the case of markers Leu 8 and Arg C4 (near 0.01% of original activity); biological activity of markers Nic 38, Me+ B5, Ura 1 was 0.2-1.2% of original level; linked markers His B2, Trp C, and His B2-Trp C has 2-6% of original level. With heterologous transformation, in some cases the sensitivities of the same markers from different Bacilli were not the same. B. subtilis 168 and B. aterrimus exhibited the maximal difference. For example, the residual biological activity of the linked His B2-Trp C markers was 50 times lower in the case of B. aterrimus DNA than in B. subtilis 168 DNA, after treatment with Eco R1. Probably, the difference was dependent of the different localization of the restricting sites on the chromosome in the closely related species of Bacilli. In the case of E. coli, only homologous transformation was studied. The transforming activity of E. coli DNA treated with Eco R1 disappeared entirely for Pro A2 marker, didn't change for His 4 and Thr 1 markers, and increased 2-5 times for Leu 6 and Arg E3 markers. The unusual result of an increased transforming activity may be caused by the reduction of molecular size of DNA fragments to the optimal for E. coli transformation.

REPRESSION CONTROL BY HOMOLOGOUS AND HETEROLOGOUS REPRESSION SYSTEM OF DIFFERENT BACTERIAL GENERA, AND THE PLASMIDS OBTAINED FROM *Bacillus subtilis* GROUP

K. Sakaguchi, K. Nagahari, T. Tanaka, M. Kuroda and T. Koshikawa
Mitsubishi-Kasei Institute of Life Sciences, Machida-shi, Tokyo, Japan

The gene dosage effect on tryptophan synthesizing enzymes which are under the repression control was investigated by constructing four composite plasmids of varing copy numbers. pSC101-trp (Copy number 4.2), RSF1010-trp (11.2), RSF2124-trp (11.9) and RP4-trp (1.6) were constructed in vitro and the tryptophan synthetase activities in crude extracts of cells containing each composite plasmid were assayed under repressed condition with the addition of 100 μg/ml tryptophan in medium or under derepressed condition with the addition of 10 μg/ml indolylacrylic acid. The results showed that the repression system of host cells affects more strongly the expression of genes than the effect of the number of genes, however, the cells containing higher gene dosage consistently exhibit more formation of the enzymes.

RP4-trp plasmid which comprises whole *E. coli* tryptophan operon was transferred into *Pseudomonas aeruginosa* trp^- cells. The activities of anthranilate synthetase and tryptophan synthetase were assayed on the crude extract of the *Pseudomonas* cells. Both were fully expressed even in the presence of 100 μg tryptophan per ml in the culture medium, showing the impotency of *Pseudomonas* tryptophan repression system on *E. coli* tryptophan operon.

Bacillus subtilis chromosomal DNA was cleaved by EcoRI endonuclease, ligated with RSF2124 plasmid DNA and introduced into *E. coli* C600$r^-m^-leu^-thr^-$ cells, resulting in leu^+ clones. The plasmid DNA was obtained from the cloned cells, chopped by EcoRI enzyme, and introduced into *Bacillus subtilis* RM125$r^-m^-leu^-$ cells, giving rise to leu^+ transformants. These experiments prove the leucine gene originated actually from *Bacillus subtilis* chromosomal DNA. The three times enhancement of isopropylmalate synthetase activity belonging within leucine operon was observed in the crude extract.

Four new plasmids were isolated from *Bacillus subtilis*. They were characterized by the cleavage site by EcoRI, Hind III and Bam NI enzymes, by their molecular weight, and classified into three different species.

RECOMBINATION BETWEEN BACTERIAL PLASMIDS LEADING TO THE FORMATION OF PLASMID MULTIMERS

John Bedbrook and Fred Ausubel
Biological Laboratories, Harvard University

We report here the formation of plasmid multimers in *E. coli*. Our experiments show that pMB9 DNA (and other plasmid DNAs) exists as various discrete sizes in *rec*$^+$ hosts. Electron microscopic and restriction enzyme analysis indicates that the larger forms of pMB9 are covalently closed circular molecules composed of tandem repeats of the smallest size pMB9 DNA. To determine the role of the host recombination genes in plasmid multimer formation, we transformed *rec*$^-$ or *rec*$^+$ strains with plasmid molecules of a single size, isolated supercoiled DNA from the transformants, and fractionated the isolated DNA on agarose gels where plasmid multimers migrate as discrete bands. These experiments show that: 1) Multimer formation occurs at high frequency in *rec*$^+$, *rec*B$^-$C$^-$, and *rec*F$^-$ hosts. 2) Multimer formation is not detected in *rec*A$^-$ hosts and occurs at reduced frequency in *rec*B$^-$C$^-$F$^-$ hosts. 3) Multimers propagate themselves stably in *rec*A$^-$ hosts and multimer formation is reversible in *rec*$^+$ hosts. 4) pMB9 forms hybrids with pML21, a plasmid which contains a limited region of homology with pMB9, in *rec*$^+$ but not in *rec*$^-$ hosts. We conclude that plasmid multimers arise by a single reciprocal recombination event occuring at regions of homology between plasmid molecules.

PLASMIDS CARRYING THE λi^{434}cro AND cII GENES EXPRESS SPECIFIC DNA BINDING ACTIVITY

B. Gronenborn and J. Messing
Institut für Genetik der Universität zu Köln and Max-Planck-Institut für Biochemie, Martinsried bei München

A purified 1.1 Mdt Eco RI fragment from phage λi^{434}tsSsus7 which covers part of the 434 cI, the 434 cro and cII genes has been inserted into plasmid pVH 51 (1) by in vitro ligation. Hybrid plasmids (pMG 2108, 2114) are able to complement the cII2002 mutation of the heteroimmune phage λ and reduce the plating efficiency of λi^{434}c to less than 10^{-4}. Fractionation of crude extracts of pMG 2108 containing cells on phosphocellulose yields DNA binding activities specific for λ and 434 DNA in the presence of a 400 fold excess of chickenblood DNA.

REFERENCES

(1) Hershfield et al. Proc. Nat. Acad. Sci. USA 71, 3455 (1974)

This work was supported by Deutsche Forschungsgemeinschaft.

THE FILAMENTOUS COLIPHAGE M13 AS A CLONING VEHICLE. INSERTION OF A HIND II FRAGMENT OF THE LAC REGULATORY REGION IN M13 RF IN VITRO.

J. Messing and B. Gronenborn
Max-Planck-Institut für Biochemie, Martinsried b. München and Institut für Genetik der Universität zu Köln

pSF 2124 was used to clone a Eco RI fragment of λ transducing phage, in which the lac regulatory region is fused to λ immunity (1,2). From this Hybrid pMG 1107 a Hind II fragment carrying the lac regulatory region and the information for the α-peptide of the ß-galactosidase was prepared by lac repressor binding (3). The replicative form of M13 has been linearized by limited digestion with the restriction endonuclease Bsu I, which cleaves the molecule at ten different sites (4). After purification the RF III linear molecules were joined by T4 DNA ligase to the Hind II fragment carrying the lac regulatory region. The mixture was used to transform a host of M13 capable to indicate α-complementation (5). Under these conditions a hybrid phage produces a blue plaque (2). Stable hybrids have been purified and used to isolate supercoiled RF DNA. Retransformation yields blue plaques (i.e. expression of the α-peptide) within an efficiency of more than 99.9%.

REFERENCES

(1) Messing and Gronenborn, unpublished results
(2) Gronenborn Molec. gen. genet. in press (1976)
(3) Landy et al. Molec. gen. genet. 133, 273 (1974)
(4) van den Hondel et al. Eur. J. Biochem. 53, 559 (1975)
(5) Ullman et al. J. Mol. Biol. 24, 339 (1967)

This work was supported by Deutsche Forschungsgemeinschaft

GENERAL METHODS FOR INSERTING SPECIFIC DNA SEQUENCES IN CLONING VEHICLES, Ray Wu, Chander P. Bahl and Saran A. Narang*. Section of Biochemistry, Molecular and Cell Biology, Cornell University, Ithaca, NY 14853, and *Division of Biological Sciences, National Research Council of Canada, Ottawa, Canada.

We have developed two general methods to introduce any double-stranded DNA molecule into cloning vehicles at different restriction endonuclease sites. In the first method, a chemically synthesized decadeoxyribonucleotide duplex, containing a specific restriction endonuclease sequence (e.g. *Bam*I, *Hind*III, *Eco*RI, *Pst*I) is joined by DNA ligase to both ends of the DNA to be cloned. The resulting new duplex DNA is cut by the same restriction endonuclease to generate the cohesive ends. It is then inserted into the restriction endonuclease cleavage site of the cloning vehicle (e.g. plasmid pMB9). The insertion of a *lac* operator into the *Bam*I and *Hind*III sites of pMB9 plasmid has been reported (Bahl, *et al.*, *Gene*, in press).

In the second method, we utilize the *lac* operator with different restriction endonuclease recognition sequences as prepared above as a means for screening any DNA molecule (X) to be cloned. The DNA molecule (X) is first joined to the *lac* operator and then inserted into the plasmid for cloning. The presence of the *lac* operator allows for an easy screening for cells harboring hybrid plasmids which makes the cells constitutive for β-galactosidase. In this way, the presence of *lac* operator and thus DNA molecule (X) in a hybrid clone can be readily detected.

The specific cloning of a *lac* operator at the *Eco*RI site has been reported recently (Marians *et al*; Heyneker *et al*; *Nature*, in press). In developing both methods, in the present work, *lac* operator was used as a convenient source of well-defined DNA in the first method, and as an indicator for screening in the second method. These two general methods should be useful for the cloning of a variety of DNA molecules and thus will facilitate solving of different problems in molecular biology.

ENZYMATIC SYNTHESIS OF RABBIT GLOBIN GENES: COMPARISON OF T4 DNA POLYMERASE AND RNA-DIRECTED DNA POLYMERASE.

Leslie J. Krueger and W. French Anderson. Molecular Hematology Branch, NHLBI, NIH, Bethesda, Maryland 20014

A comparative study of avian myeloblastosis virus DNA polymerase (reverse transcriptase) and DNA polymerase purified from *E. coli* cells infected by T4 bacteriophage was undertaken to determine the efficiencies of these enzymes in synthesizing the second strand DNA (sDNA) to rabbit globin complementary DNA (cDNA). We have evaluated the effect of different reaction conditions on the self-primed polymerization of deoxynucleotides into sDNA (initially described by Efstratiadis *et al*[1] for *E. coli* DNA polymerase I) and have also analyzed the reaction products for the length of the sDNA-cDNA polymer.

Poly(A) associated RNA was prepared from salt washed rabbit reticulocyte polysomes which were passed over a 5 gm column of oligo(dT)-cellulose. The RNA was then loaded onto a 2.5 cm x 100 cm Sepharose 4BCL column and biologically active globin mRNA was pooled and shown to be pure 10S material by 6% polyacrylamide electrophoresis in 98% formamide. Full sized cDNA was prepared by the method of Weiss *et al.*[2] except actinomycin D was eliminated from the reaction cocktail. The cDNA was sized on a 5-30% alkaline sucrose gradient, ethanol precipitated without carrier and collected by ultracentrifugation.

Conditions appropriate for polymerization were initially evaluated using salmon sperm nicked DNA as the template. Using these as standard conditions, an enzyme titration curve was generated on salmon sperm DNA such that reverse transcriptase and T4 polymerase could be compared directly. Reverse transcriptase demonstrated maximal sDNA synthesis at a ratio of 4 μg cDNA to 3 units enzyme activity; however T4 polymerase was more efficient and more effective in synthesizing full sized sDNA. In the T4 polymerase reaction, greater than 50% of the template was copied when assayed by trichloroacetic acid precipitation of the polymerized deoxynucleotide. The size distribution evaluated by composite acrylamide-agarose gels in 98% formamide and confirmed by electron microscopy demonstrated that approximately 70% of the molecules ranged from approximately 500 to 650 base pairs. Amplification of these genes in prokaryotes is being attempted in order to access their quality.

(1) A. Efstratiadis, F.C. Kafatos, A.M. Maxam, and T. Maniatis *Cell* 1: 279-288 (1976).
(2) G.B. Weiss, G.N. Wilson, A.W. Steggles, and W.F. Anderson *J. Biol. Chem.* 251: 3425-3431 (1976).

ORGANIZATION OF BOMBYX MORI rDNA. R. F. Manning and L. Patrick Gage. Roche Institute of Molecular Biology, Nutley, N. J. 07110.

The organization of the ribosomal RNA genes (rDNA) of the silkworm, *Bombyx mori* has been examined through the use of restriction endonucleases. rDNA restriction fragments from total *B. mori* DNA were detected and sized by agarose gel electrophoresis and hybridization with *B. mori* ^{125}I-rRNA according to the technique of Southern [J. Mol. Biol. (1975) 98, 503]. After digestion with either EcoRI, HindIII or HpaI, the rDNA electrophoresed as a single band with a molecular weight of 6.8 x 10^6d. Hence, the rRNA genes, which number about 250 per genome, must be organized in a tandem array with this repeat length. Several enzymes, among them BamHI, HindII and HaeIII, cleave each rDNA repeat at multiple sites, and SalI does not cleave the array at all. Analysis of digests made with two or more endonucleases, and hybridization with 18S or 28S ^{125}I-rRNA has allowed the mapping of their cleavage sites with respect to the rRNA genes. This map is shown below. We have found no evidence for major length heterogeneity of the tandem rDNA repeats due to "insertions" like those reported for *Drosophila melanogaster* (Wellauer and Dawid, in press), and we are currently examining the length of the spacer sequences for micro-heterogeneity.

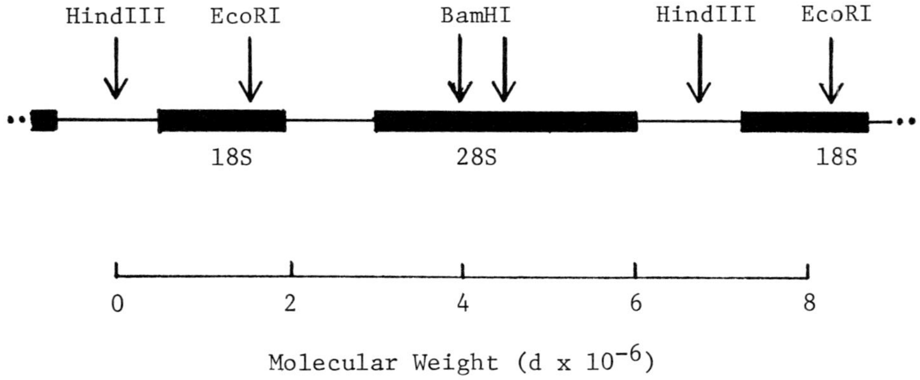

CLONING AND CHARACTERIZATION OF YEAST DNA

Kamalendu Nath and Arthur P. Bollon
University of Texas Health Science Center
Dallas, Texas

Yeast DNA has been cloned by using the plasmid PBR 312. This plasmid which was obtained from Dr. Herb Boyer contains two resistance genes: ampicillin (Amp) and tetracycline (Tet.). The Tet gene contains a Hind III sequence and is therefore inactivated when foreign DNA is inserted at the Hind III site. Yeast DNA and PBR 312 DNA were treated with Hind III, a mixture of both DNA species were ligated and the ligated mixture was transformed into a strain which was restriction minus (r^-), modification minus (m^-), Tet^s, and Amp^s. For 1 μg of ligated DNA we obtained 5 x 10^5 transformants, namely, clones which are Amp^R. Of the transformed clones analyzed 35% of such transformants are Amp^R Tet^S implying that they contain cloned yeast DNA. Hence 10^4 clones containing yeast DNA have been isolated. Since we have calculated the average size of the yeast Hind III generated DNA fragments as 7 x 10^6 daltons, we have cloned about 70 times the size of the yeast genome (yeast genome is about 1.2 x 10^{10} daltons). Using the in situ hybridization technique developed by Hogness, we have determined the frequency of transformed clones which contain ribosomal DNA sequences. Of the 10^4 transformants which are Amp^R Tet^s, 3% of the clones hybridized to total ribosomal RNA. Such a frequency of ribosomal genes is consistant with the frequency of ribosomal DNA present in the yeast genome which is 2.5%. This may imply that the cloned yeast DNA represents a random distribution of yeast DNA. We have cloned and characterized the various ribosomal 5S, 25S, 5.7S and 18S DNA components and have ordered the genes. In addition to using RNA as a probe for characterizing the cloned yeast DNA fragments we are using antibody to specific enzymes and testing for yeast gene expression in the bacterial clones. Positive selection of yeast structural genes in bacteria is also to be studied. Details of the characterization of the cloned yeast DNA fragments will be presented.

TRANSFORMATION PROCEDURE TO *E. COLI* χ 1776 STRAIN

M. Inoue and R. Curtiss III
University of Alabama in Birmingham, Birmingham, Alabama

It is difficult to obtain efficient transformation with the χ 1776 strain because χ 1776 is much more sensitive than other *E. coli* to ionic detergents, bile salts, temperature and many drugs. Also the χ 1776 strain is easily lysed after washing with $CaCl_2$.

We have been studying several conditions for transformation to χ 1776, such as temperature, pH, $CaCl_2$ concentrations and cell concentration effects. From these results we have found the following transformation procedures. The overnight culture is diluted 10-fold into L-broth with Dap (100 μg/ml) and Thymidine (40 μg/ml) and incubated for 3.5 hr at 37C with shaking. The 20 ml cell cultures are then centrifuged at room temperature in a sterile 50 ml polypropylene tube. The pellet is gently resuspended in 10 ml of 100 mM NaCl in H_2O and centrifuged for 10 min at room temperature. Then the pellet is resuspended in 10 ml of 10 mM Tris (pH 8.0 at room temp.)-0.8% NaCl-75 mM $CaCl_2$ and kept at room temp. for 20 min. Cells are then centrifuged again at room temp. for 10 min. and gently resuspended in 2 ml of the same buffer. Cells are put into the small glass tube and chilled on ice for 2 min. before mixing the DNA. 0.2 ml chilled cells and 0.1 ml DNA (in 0.8% NaCl-20 mM Tris pH 8.0 at room temp.) mixture is kept on ice for 20 min. then heat shocked at 42C for 1 min. and kept on ice for 10 min. The transformation mixture is plated on EMB or PA agar plates supplemented with Dap and Thymidine and 12.5 μg/ml TC. Plates are incubated at 37C for 3 days.

RESTRICTION ENDONUCLEASES: PROTECTION OF PARTICULAR RECOGNITION SITES WITH ANTIBIOTICS

O.L. Polianovski, V.V. Nosikov, A.L. Zhuze, E.A. Braga
Institute of Molecular Biology, Academy of Sciences of the U.S.S.R., Moscow B-312, U.S.S.R.

The antibiotics distamycin A and actinomycin D interact with double-stranded DNA forming noncovalently bound complexes. Distamycin preferentially interacts with A/T-rich regions whereas actinomycin binds G/C-rich regions of DNA. The actions of antibiotics were compared on endo R· Eco RI, Eco RII, Hind III, Hpa I, Hpa II and Sma I. It was shown that antibiotic action depends on the nucleotide sequences of the recognition sites and their nearby environment. At appropriate distamycin or actinomycin concentrations only particular recognition sites on DNA are available for the restriction enzymes. This phenomenon results in the appearance of larger DNA fragments. It allows us to obtain a set of overlapping DNA fragments. After removal of the antibiotics the DNA fragments may be used for further experiments.

1. Braga, E.A., Nosikov, V.V., Tanjashin, V.I., Zhuze, A.L., Polianovski, O.L. Doklady Akad. Nauk USSR 225 (1975) 707.

2. Nosikov, V.V., Braga, E.A., Karlishev, A.V., Zhuze, A.L. Polianovski, O.L. Nucl. Acid Res., 3 (1976) 2293.

3. Braga, E.A., Nosikov, V.V., Polianovski, O.L. FEBS Letters, 70 (1976) 91.

A
B 7
C 8
D 9
E 0
F 1
G 2
H 3
I 4
J 5